Before the Age of Dinosaurs in Virginia and Nearby States

BEFORE
THE AGE OF
DINOSAURS
IN VIRGINIA AND NEARBY STATES

ROBERT E. WEEMS, PH.D.

GARY J. GRIMSLEY

BELLE ISLE BOOKS
www.belleislebooks.com

ISBN (Paperback): 978-1-962416-52-8
ISBN (eBook): 978-1-962416-53-5
Library of Congress Control Number: 2024914620

Designed by Sami Langston
Project managed by Casey Cornish

Printed in the United States of America

Published by
Belle Isle Books (an imprint of Brandylane Publishers, Inc.)
5 S. 1st Street
Richmond, Virginia 23219

BELLE ISLE BOOKS
www.belleislebooks.com

belleislebooks.com | brandylanepublishers.com

ACKNOWLEDGMENTS

The authors gratefully thank Casey Cornish at Brandylane Publishers for her extensive help in getting our volume ready for publication.

Contents

CHAPTER

1

THE OLDEST ROCKS IN THE VIRGINIA REGION

"The Age of Dinosaurs in Virginia and Nearby States," published previously, covers the known history of the Mesozoic Era in Virginia and nearby regions. Yet even by the beginning of the Age of Dinosaurs, planet Earth had already existed for a very long time. The Earth and the other planets in our solar system formed about 4,600 million years ago (Ma).[1] This initial event subsequently was followed by the formation of the oceans of the world,[2] the appearance of life on Earth, [3] and the evolution of the Earth's atmosphere until it became at least similar to what we see today. [4] Indeed, so much time passed during this early part of our planet's history that, by the beginning of the age of dinosaurs around 252 Ma, planet Earth already had attained about 95% of its present age. Although most of Earth's history had already come and gone by the time that the Age of Dinosaurs even began, unfortunately, we know relatively little about the earliest part of the prehistory of Virginia. Earth's rocks have been recycled repeatedly by plate tectonics ever since our planet first formed.[5] This recycling has caused the original crust of the Earth to be destroyed and reformed repeatedly, creating the rocks that we see today. Even though Earth's oldest rocks and minerals have been recycled and not destroyed, the process of recycling them has been destructive enough to erase most of the earliest history they once contained.

In the Virginia region, the oldest rocks discovered so far formed during the Grenville mountain building event, which occurred about 1.2 to 1.0 billion years ago.[6] Even though these oldest rocks are rather spectacular to see, especially where they are exposed at the heights of the Blue Ridge Mountains in Virginia (Figure 1), it is important to keep in mind that most of them have been altered through several cycles of mountain building during the course of their history. Additionally, these oldest rocks in Virginia actually formed deep beneath what then was the surface of the Earth. For this reason, there is no direct evidence remaining within them of any of the fossil creatures that were living at that long-ago time or even of the conditions at Earth's surface back then. Based mostly on what we know from other parts of the world, it seems likely that Virginia back then was in the heart of a Himalayan-style mountain belt with very high peaks and no land-based life to be found within what then was a desolate upland landscape. Beyond this broad perspective, we can say nothing more about the conditions that prevailed in Virginia during that very ancient time. Any rocks that may have once retained evidence of the life and climate in Virginia at that time were eroded away or utterly deformed long ago so that the story of this very early time in the history of Virginia can only be surmised from better-preserved rocks found elsewhere in the world.[7]

In Virginia today, the oldest known rocks that contain any clear evidence of what conditions were like at the surface of the Earth belong to the Cryogenian Period of the Neoproterozoic Era, which happened much later, around 700 Ma (Figure 2). At that time, well before the dinosaurs, Virginia was an utterly different place than it is today. First of all, it was not just one place. Back then, both of the major pieces of Virginia were probably located in the southern hemisphere far to the south of where they are now (Figure 3). What we know about the ancient locations of the modern pieces of Virginia has been discovered because the least deformed rocks of Proterozoic age in North America retain something known as "remanent magnetism," which tells us the latitude (but not the longitude) of the rocks that formed at that distant time.[8]

In the Cryogenian Period, western Virginia was a part of North America as it is today, but eastern Virginia was part of an entirely different ancient continent called Gondwana (Figure 1). Between Neoproterozoic North America (including modern western Virginia) and Gondwana (including modern eastern Virginia), there was an island arc now called the Carolina Terrane. Geologically, it was similar to the Lesser Antilles of today, overlying a plate tectonic subduction zone that had built up through volcanic eruptions over a period of 30 or 40 million years before it was thrust against the eastern edge of North America around 440 Ma.[9] This collision caused what then was the eastern edge of Virginia and nearby states to be thrust up and over the more inland portions of the continent, creating the Blue Ridge and American Piedmont regions of today. After this early Paleozoic addition to the eastern border of Virginia, Gondwana (which includes modern eastern Piedmont and Coastal Plain regions) came crashing into North America and added the eastern part of Virginia to the state around 330 Ma. The end result of this prolonged process was the creation of the ancestral Appalachian Mountains, which today are the deeply eroded roots of the landscapes of central and western Virginia.

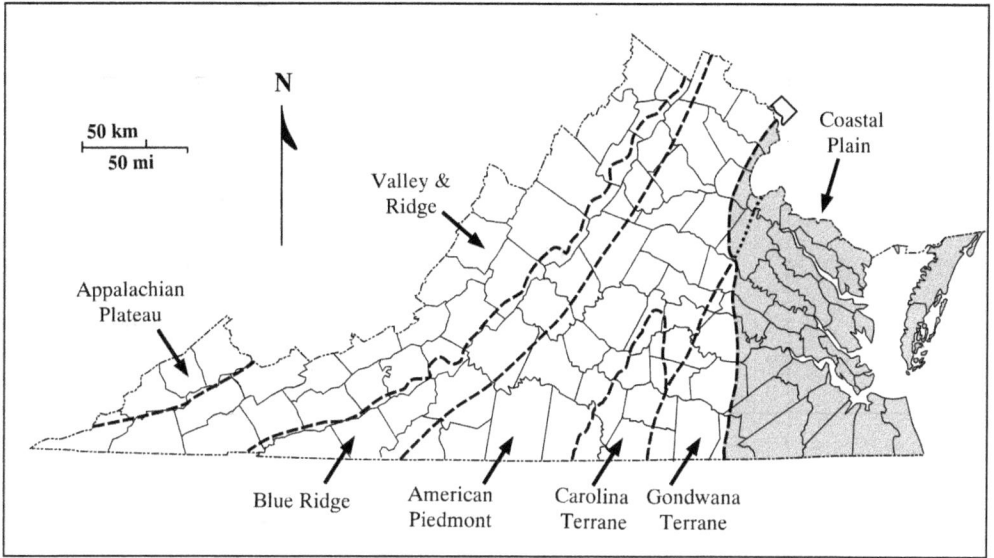

Figure 1 — *The major geologic regions of Virginia. At the beginning of the Paleozoic Era, only the parts of Virginia west of the Carolina and Gondwana Terranes belonged to North America. It was not until middle and late Paleozoic time that the Carolina and Gondwana Terranes were shoved into the present eastern edge of North America, forming the ancestral Appalachian Mountains. The Coastal Plain on the east side of the state is directly underlain today by much younger Mesozoic and Cenozoic deposits, but buried beneath them are rocks that were once mostly part of the Gondwana Terrane (east of the dotted line). See Figure 3 for the most ancient estimated locations of the eastern and western portions of Virginia. The ancient location of the Carolina Terrane remains uncertain but was probably near what is now South America (also called Amazonia). See Appendix 2 for the source of this picture.*

EON	ERA	DATE (Ma)	PERIOD
LOWER PHANEROZOIC	PALEOZOIC	252	PERMIAN
		299	PENNSYLVANIAN
		323	MISSISSIPPIAN
		359	DEVONIAN
		419	SILURIAN
		444	ORDOVICIAN
		485	CAMBRIAN
		541	
UPPER PROTEROZOIC	NEOPROTEROZOIC		EDIACARAN
		635	CRYOGENIAN
		720	
			TONIAN
		1000	

Figure 2 — A geologic time scale that includes the earliest part of Virginia's known sedimentary history. The oldest sedimentary rocks known to still exist in Virginia belong to the Cryogenian Period of the Neoproterozoic Era and are about 700 million years old. Elsewhere in North America, there still exist a few areas with sedimentary rocks as old as 3,900 million years.[7] In older literature, rocks older than the Cambrian Period often were simply described as being "Precambrian." See Appendix 2 for the source of this picture.

CHAPTER

2

THE CRYOGENIAN:
EVIDENCE OF AN ANCIENT ICE AGE IN VIRGINIA

It was during the Cryogenian Period (Figure 2) that surficial deposits formed in western Virginia that still remain, giving us a direct window into the world in which western Virginia then existed. From these deposits, we know that the world during the Cryogenian Period was profoundly different climatically from today's world. From 720 to 635 Ma, the Earth experienced two massive glaciations that were far more severe and extensive than any that are known to have occurred on Earth before or since then, including our "recent" Pleistocene glacial age. During this very ancient period, appropriately named the Cryogenian, most or perhaps all of the surface of the Earth's oceans froze over during several intervals of time.

It was during this interval of time that the oldest known unmetamorphosed sedimentary rocks were forming that are still preserved in Virginia. These surficial sedimentary rocks are found in the central part of the Blue Ridge Mountains southwest of Charlottesville and in the southern part of the Blue Ridge Mountains near Mount Rogers (Figure 4). These strata include beds that contain fragments of rock bearing the telltale markings and texture of glacial deposits. [10] These deposits apparently formed at or near the equator back then, based on their paleomagnetic latitude. This is one of the reasons for believing that the glaciers that formed back in Cryogenian time likely covered almost the entire Earth. This means that, even though the western Virginia landscape was latitudinally tropical during the Cryogenian Period, it was still covered with glaciers down to sea level during this earliest, fairly well-known chapter of Virginia's prehistory.[11] It is within this strange and globally glacial world that the earliest discernable chapter began in the geologic history of Virginia's landscape.

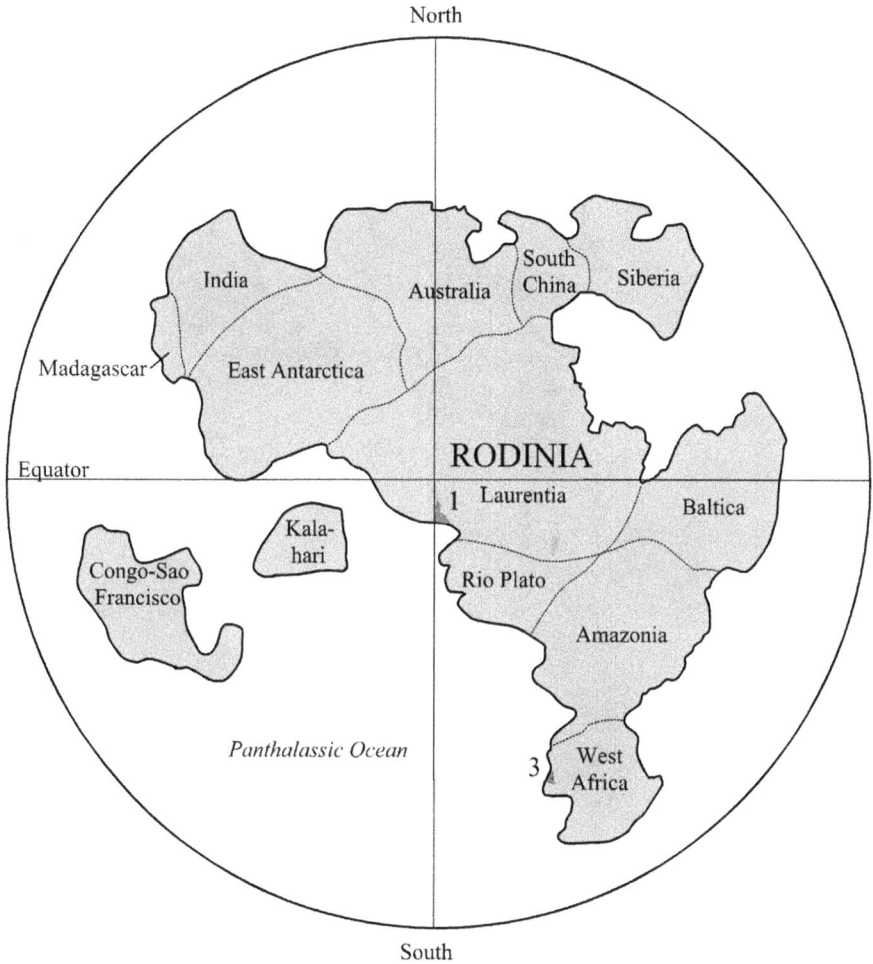

Cryogenian
(ca. 720 Ma)

Figure 3 — Location of western Virginia (labeled 1, dark gray) and eastern Virginia (labeled 3, dark gray) in the Cryogenian Period relative to the positions of the modern continents and continental fragments. These lands were all part of the supercontinent of Rodinia during the Cryogenian Period (720 to 635 Ma). Back then, western Virginia lay slightly south of the equator while eastern Virginia likely lay well south of the equator in a very different place. The Carolina Terrane (part 2) had not yet formed. See Appendix 2 for the source of this picture.

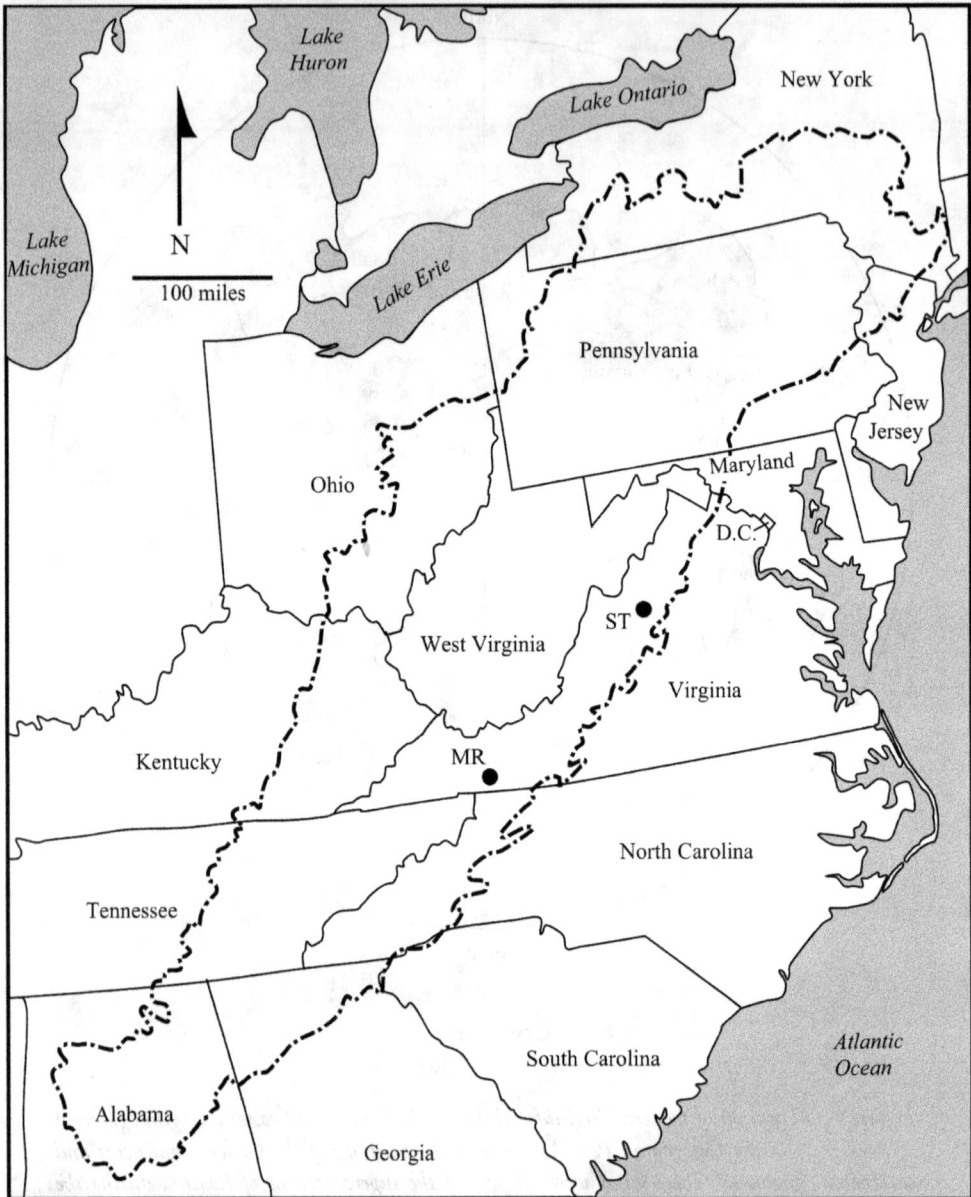

Figure 4 — The location of Cryogenian glacial deposits in Virginia. MR = Mount Rogers, the highest mountain in Virginia. ST = Sharp Top Mountain, located 20 km southwest of Charlottesville, Virginia. These areas have the oldest known glacial and sedimentary rocks yet found in the state of Virginia. The modern Appalachian region is enclosed by a dotted and dashed line. See Appendix 2 for the source of this picture.

Figure 5 — Outcrop of the Konnarock Formation near Mount Rogers, showing the ancient glacial rocks that formed over 700 Ma in this area. Scale is in centimeters. Even though this region was near the equator back then, glaciers were so extensive that they covered most of the surface of the Earth. See Appendix 2 for the source of this photograph.

CHAPTER

3

THE EDIACARAN:

THE EARLIEST KNOWN MULTICELLULAR LIFE

Once the Cryogenian Period ended, the Earth began to warm again and life began to adapt to its new and warmer world. During the Ediacaran Period, between 635 and 541 Ma, multicellular life began to develop in shallow and now ice-free seas. Nearly all the fossils found in rocks that formed during this interval of time are very different creatures from the animals and plants living today, so much so that many of them cannot be assigned to any existing group of animals or plants still living. So far, no Ediacaran fossils have been found in Virginia, but a few have been found in the central part of the Carolina Terrane in North Carolina (Figure 6). Carolina Terrane rocks extend up into Virginia (Figure 1), so it is possible that some examples of fossils from this time interval will eventually be found here.

The Ediacaran fauna preserved in North Carolina (Figure 7) includes several kinds of fossils that have been named *Pteridinium carolinaensis*, cf. *Swartpuntia* sp., *Sekwia excentrica*, and *Aspidella* cf. *A. terranovica*. Elsewhere, trace fossils are found that also indicate the former presence of soft-bodied crawling animals that then lived in this world. These very ancient creatures were quite different from animals living today and it is not at all clear if they are related to any kinds of modern animals or whether they are entirely extinct. It is not until much later, at the beginning of the Cambrian Period, that creatures finally appeared that began to closely resemble some of the creatures that are still alive today. A recent study indicates that the Ediacaran fauna underwent a major extinction event near the end of the Ediacaran Period. This was likely caused by a global catastrophe that seriously reduced the abundance of oxygen in the seas of that time.[12]

The part of Virginia in which these creatures probably lived was part of what today is known as the Carolina Terrane or Carolinia. Its identity is based on younger Cambrian fossil organisms found farther to the south in this same region (Figure 6).[13] No Ediacaran fossils are known from the western or eastern parts of modern Virginia, and it is unlikely that either of these two regions was below sea level back then. Therefore, even though the Cryogenian fauna that has been found around the world was fairly uniform back then,[14] the eastern and western parts of modern Virginia were likely above sea level (Figure 8) and so were barren of any multicellular life.

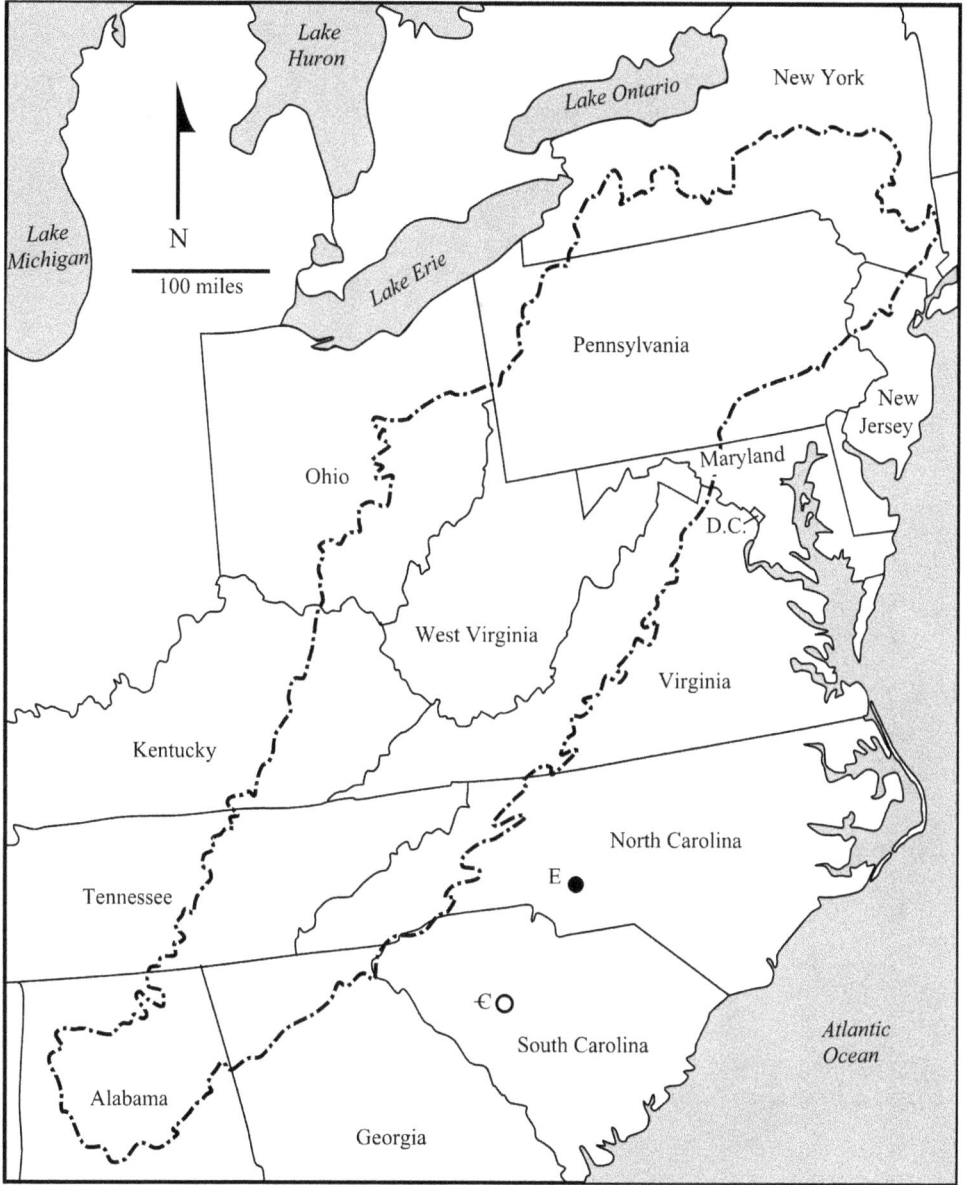

Figure 6 — Location of Ediacaran fossils (E) found in the Carolina Terrane of North Carolina. A somewhat younger Cambrian locality, found in this same terrane in South Carolina (open circle), indicates that it was not a part of North America at that time. The modern Appalachian region is enclosed by a dotted and dashed line. See Appendix 2 for the source of this figure.

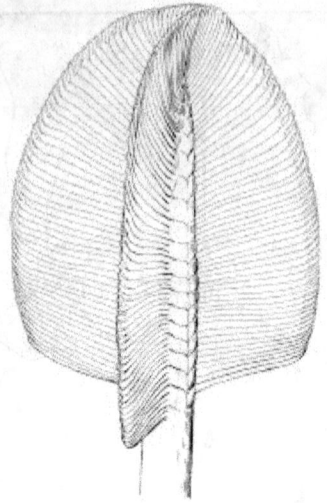

Pteridinium carolinaensis   *Swartpuntia* sp.

Figure 7 — Two Ediacaran animals known from the Carolina Terrane of North Carolina. See Appendix 2 for the sources of these pictures.

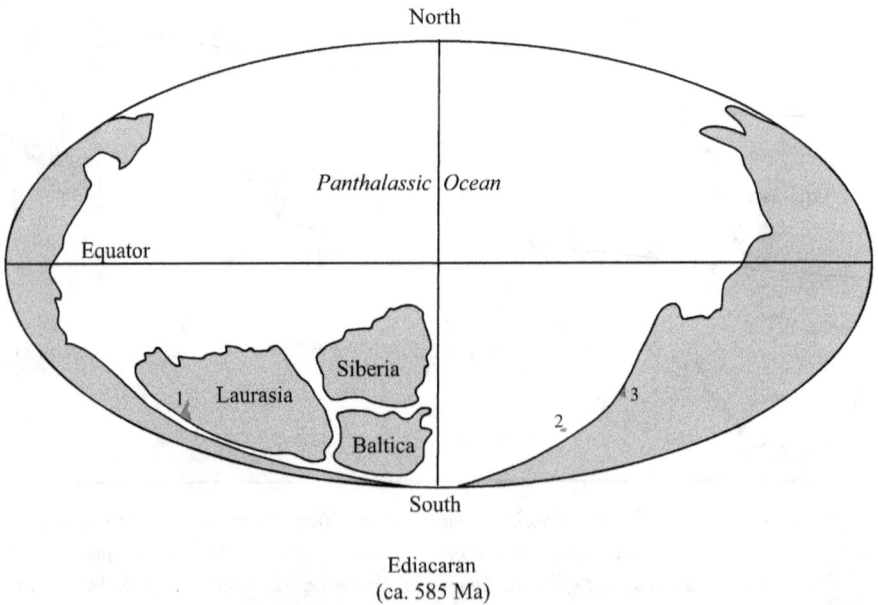

Ediacaran
(ca. 585 Ma)

Figure 8 — Location of the parts of Virginia (shown as numbered and dark gray areas) during the Ediacaran Period. Carolinia (or the Carolina Terrane, numbered 2) formed at this time as part of an island arc off the western coast of what is now South America. Note that all parts of Virginia lay between 30 and 70 degrees south latitude and were far apart from each other in longitude. See Appendix 2 for the source of this picture.

CHAPTER

4

THE CAMBRIAN:
EARLY FISH RELATIVES APPEAR IN THE APPALACHIAN REGION

After the Cryogenian Period, the record of Paleozoic life in Virginia is known exclusively from the western parts of the state and from states that today border this region. Rocks that formed at this time in the Carolinian and Gondwana terranes (Figure 1) were later strongly heated and metamorphosed when they were thrust into and became attached to the rest of Virginia. As a result of this, nothing remains that gives us any clue about what life was like at that time in those regions of the eastern part of our state. The one exception to this is the occurrence of Cambrian fossils in the Carolina Terrane in South Carolina. These fossils, especially the trilobites, clearly indicate that this terrane was far away from western Virginia during the Cambrian and not at all closely connected to the Cambrian life whose remains are found today in western Virginia.[13]

At the beginning of the Cambrian, about 541 million years ago, modern types of animal life began to appear on Earth. Animals already existed before this time, but they had no hard parts composed of calcium carbonate, calcium phosphate, or silica. Something changed in the environment of the Earth at about this time, and this change allowed animals to start concentrating hard materials to provide protection and cover for their bodies.[15] This created a major shift toward the protection of life on Earth, resulting in an "arms race" among creatures to shield themselves from predators and their environments.

A great many of Earth's creatures quickly took up this arms race and, with the dawn of the Cambrian Period, a great many invertebrate groups began to become common in the fossil record. By protecting their bodies with hard parts, they also became predisposed to become preserved as fossils. At the beginning of this new age, the trilobites and brachiopods were especially successful and abundant

(Figure 9), and they would remain prominent in the western Virginia fossil record for all of the succeeding Paleozoic Era. Notably, fish are one group that is prominent today but remains undocumented within the Cambrian record of Virginia. Very early fish-like creatures do appear in the Early Cambrian strata of China,[16] but so far we have no record of them in North America. The oldest North American fossil fish (*Metaspriggina*) is known from the Middle Cambrian Burgess Shale in British Columbia, Vermont, and also in Pennsylvania (Figure 10). It was a soft-bodied animal that has no hard skeletal structures, though it had gills and a very fish-like body.[17] *Metaspriggina* (Figure 11) probably also lived here in Virginia, but it has not been found here yet because it was entirely soft-bodied and had no hard parts capable of ready fossilization.

Another fossil group that first appears in the Late Cambrian of Virginia is conodonts, which are represented by tooth-like elements of animals that were a primitive type of chordate creature (Figure 12). This is a group that apparently was close in ancestry to the group from which fish evolved. There has been a great deal of controversy over the years as to whether these creatures can be fairly linked to vertebrates.[18] Current thinking is that they are indeed the closest known relatives to vertebrates and are found in the fossil record as far back in time as the Late Cambrian. So far, twelve types of conodonts have been found in the Late Cambrian of Virginia (Table 1), but it is likely that many of them represent different structures within the same biological creature. Six kinds of conodont shapes are typically found within each conodont animal, so the twelve kinds of conodonts that are known from the Late Cambrian may only represent two different kinds of animals. For this reason, these earliest near-vertebrate fossils represent only a small fraction of the numerous kinds of animals that are known from the Cambrian seas of western Virginia.

Back in the Cambrian, Virginia consisted of three major regions

(Figure 13). These regions were all located very far south of their present position and also very far apart from each other in their longitude. Additionally, the western part of Virginia was turned clockwise about ninety degrees relative to its present position. This meant that the Appalachian seaway that covered western Virginia then was oriented in a more or less east-west direction slightly south of the equator and in a very tropical environment. The south-central part of the state was the northern end of the Carolina Terrane, which was located close to Gondwana in those days. Similarly, the eastern part of Virginia then was likely above sea level and therefore did not preserve any fossil remains of oceanic life.

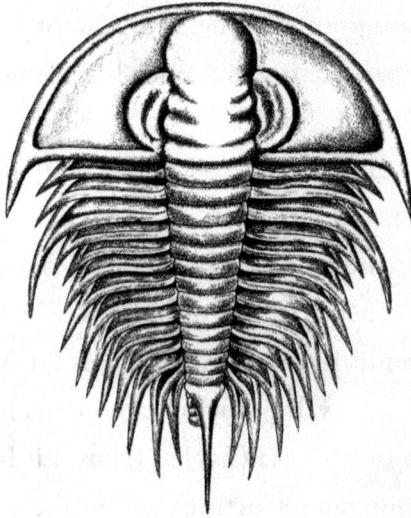

2.5 cm

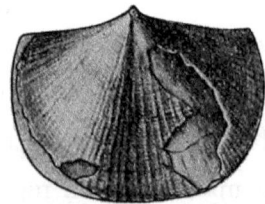

Olenellus thompsoni   *Nisusia festinata*

Figure 9 — A trilobite (left) and brachiopod (right) are examples of fossils commonly found in the Cambrian strata of Virginia. Vertebrate remains, represented by conodonts (Figure 12), are present in the Late Cambrian rocks of Virginia but are rarely found. See Appendix 2 for the sources of these pictures.

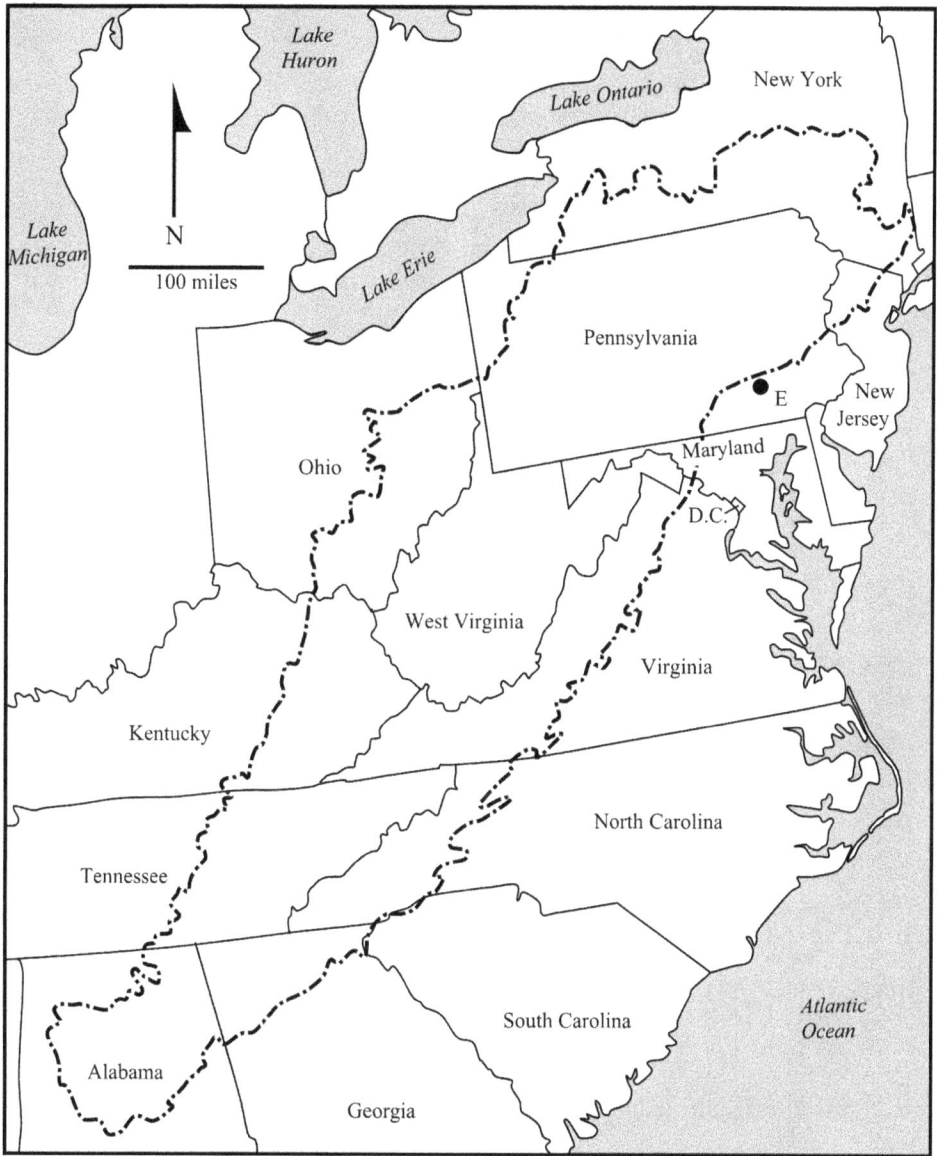

Figure 10 — Location of the Cambrian Metaspriggina *vertebrate site (E) in Pennsylvania. The modern Appalachian region is enclosed by a dotted and dashed line. See Appendix 2 for the source of this picture.*

Figure 11 — Reconstruction of the earliest known Appalachian Cambrian vertebrate, Metaspriggina, *as it likely appeared in life. This very primitive fish was only about 2 inches (5 cm) long. Gill openings, a mouth, and eyes are present, but fins have not yet appeared anywhere along its body. See Appendix 2 for the source of this picture.*

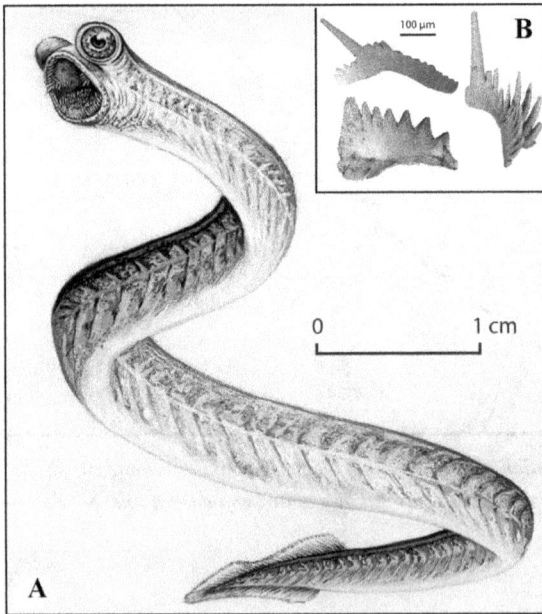

Figure 12 — (A) Reconstruction of a conodont animal (Clydagnathus) *as it likely appeared in life. This kind of creature may have been the first fish-relative to have left fairly abundant fossil remains in the Virginia region during the Late Cambrian Period. (B) Representative tooth-like conodonts that were present in the oral and throat regions of this kind of animal. See Appendix 2 for the source of this picture.*

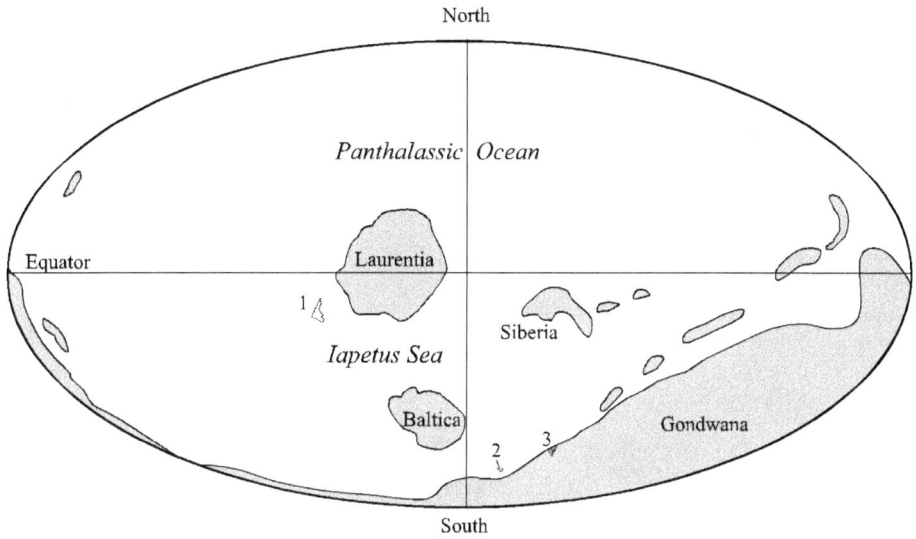

Late Cambrian
(ca. 514 Ma)

Figure 13 — Locations of the three parts of Virginia during the Late Cambrian Period. In the Late Cambrian, western Virginia (1) lay entirely beneath the Iapetus Sea, while the Carolina Terrane (2) was in the sea far to the south near Gondwana. Both regions are shown with no shading. Eastern Virginia (3) lay far to the southeast along the western edge of the continent of Gondwana and probably was then above sea level.

CHAPTER

5

THE ORDOVICIAN:

A TIME WITH VERY PRIMITIVE APPALACHIAN NEAR-VERTEBRATES

The Cambrian Period was followed by the Ordovician. In general, the Earth remained fairly stable, which allowed invertebrate life to diversify as it sought to expand its presence in the seas of the world. During this time, vertebrates also began to diversify and fill an ever-larger realm in the shallow oceans, especially toward the middle and later parts of the Ordovician. In North America, the first signs of abundant fish life began to appear in the shallow seas that then spread across the central United States and Canada, southwest of an ancient continent called Laurentia (Figure 13). Abundant small fish armor plates are particularly well-known from the Late Ordovician Harding Sandstone in Colorado and Wyoming (Figure 14), and sparse remains of these creatures also have been documented from the Whitewood Formation in South Dakota, the Arbuckle Mountains in Oklahoma, the Gull River Formation in Ontario, Canada, and the Black River Group in Quebec, Canada.[19] Perhaps armored fish remains will eventually be found here, but for now the record in the Appalachian region is blank.

It remains an interesting question why some vertebrate animals first began to get an armor coating on their bodies during the Ordovician. The best explanation so far is that these fishes were subject to predation by large carnivorous arthropods called eurypterids (Figure 15), which appeared at about this time and had very strong claws for grasping and crushing prey. This theory has been reinforced by the publication of evidence for eurypterid predation upon somewhat later Paleozoic fishes.[20]

Although true bony fish did not seem to be present during the Ordovician in the waters of the Appalachian seaway, their cousins (called the conodont animals) were relatively abundant here, probably because they lived in the upper surface layer of the open oceans where they could drift with the currents to whatever locality was producing abundant food for them to eat. The 55 types of Ordovician conodonts described from the Appalachian basin probably represent only about ten biological species that were present in the local environment for the reasons discussed in the previous chapter. These animals probably lived a life rather reminiscent of modern-day eels. Even so, their lack of a lower jaw would have rendered their appearance quite strange to anyone who might have had occasion to gaze upon these creatures as they swam in the ancient Ordovician Appalachian seaway.

While true fish seem to have been absent in the Appalachian seaway during the Ordovician, the shallow waters of the Ordovician in central North America provided habitat for at least thirteen now-extinct types of fish. Four fossil genera (*Canyonlepis, Salinalepis, Tantalepis,* and *Tezakia*) are currently considered to be shark-related fishes (known as Chondrichthyans). Other now-extinct fishes then living represented twelve other groups with varied anatomy. Of these various groups of midcontinent fishes, three went extinct at the end of the Ordovician but ten continued to survive into the Silurian.[21]

5 cm

*Figure 14 — The oldest well-known true fish (*Astraspis desiderata*) from the midcontinent region west of Virginia during the Ordovician Period. To date, no true Ordovician fish are known from Virginia, though closely related conodont animals by then were abundant here. In the Ordovician, some fossil fishes were becoming heavily armored to protect themselves from predation. See Appendix 2 for the source of this picture.*

Figure 15 — Eurypterids, such as Pterygotus, were large carnivorous arthropods that became abundant in the world's oceans during the middle of the Ordovician Period. They appeared in the geologic record at the same time that fishes began to develop external armor to protect themselves from predation. It is likely that these eurypterid predators, which could be more than five feet in length, created a major evolutionary pressure on fishes to develop protective armor. This greatly helped these fish to repel the attacks of these large predators and also made them much more likely to be preserved in the fossil record. See Appendix 2 for the source of this picture.

CHAPTER

6

THE SILURIAN:
THE OLDEST TRUE FISHES APPEAR IN VIRGINIA

The long Ordovician Period ended with a major uplift along the eastern edge of the Appalachian region, including the region now occupied by Virginia. Before the Silurian, the western Virginia region lay entirely beneath the sea and accumulated a rather steady stream of carbonate and muddy sediment that came from the regions that lay to the west of the modern Appalachians. Toward the end of the Ordovician, however, sediment for the first time began to arrive from what is today the eastern side of Virginia. This was apparently happening because the eastern edge of what then was North America began to collide with the Avalonian Terrane, which pushed the eastern edge of Paleozoic western Virginia upward. As this region rose above the sea, it began to undergo erosion and shed sediment from what was then the southern edge of Virginia northward toward the sea floor that still lay across the northern and western portions of Virginia at that time (Figure 16). Interestingly, a very similar pattern of uplift and sedimentation is occurring today along the outer edge of the South China Sea, where Taiwan is being uplifted by oceanic under-thrusting. In this modern example, Taiwan's uplifted and exposed sediment is being shed westward across the eastern South China Sea toward the Chinese mainland.[22]

This change in regional geography resulted in the earliest known occurrence of bony fish remains in the rock record of Virginia. In the farthest southwestern part of the state (Figure 17), a fish bone referable to the genus *Tremataspis* has been found in the basal Silurian Clinch Sandstone.[23] Interestingly, the closest relatives of this fish are found in Estonia in Eastern Europe, which suggests that these fish gained access to the Virginia region through the collision of Avalonia with North America during Late Ordovician and Early Silurian time.

By the Late Silurian, more fish began to appear in the Appalachian mountain region. These include cyathaspid fishes referable to the genera *Vernonaspis* and *Americaspis* (Figure 18), which are known from the Wills Creek Formation in New York, New Jersey, Pennsylvania, Maryland, and West Virginia. Somewhat later, in the Bloomsburg Formation of Pennsylvania, thelodontian fish remains (*Logania* and *Thelodus*) also appeared. All of these Silurian Appalachian fishes were jawless (agnathans) and did not yet have any obvious fins along their bodies to increase their stability in turbulent waters. In this regard, they were rather primitive for their day, because jawed fishes had already appeared in China by the beginning of the Silurian.[24]

Despite the appearance of jawless agnathan fishes in the shallow Appalachian sea, conodont animals continued to be abundant in these same waters and they were actually slightly more abundant than they had been in the Ordovician. Sixty-four taxa of conodonts have been named from these rocks, which probably represent about eleven biological species. This suggests that conodont animals were far more diverse in the Silurian than the agnathan fishes, which then were represented by only five known species. Even so, bone-bearing vertebrates were finally in the process of making an appearance in the fossil record of the Appalachian region during the Silurian Period. Their numbers and variety would markedly increase as time passed.

Also, during the Silurian, when bony fish were making their first appearance in the Virginia region, another major event in the history of life was taking place. In Massanutten Mountain, fossils have been found that document the earliest known examples of terrestrial plant life, which was beginning to take root on the land. Before the Silurian, the terrestrial realm was essentially devoid of life. With the advent of the Silurian, however, colonization of land areas of the world was starting to get underway.[25] This earliest terrestrial life was initially confined to marshy areas close to permanent water, but even so, it was still the beginning of a major change in the history of life on Earth, which until the Silurian had been restricted to the world's oceans.

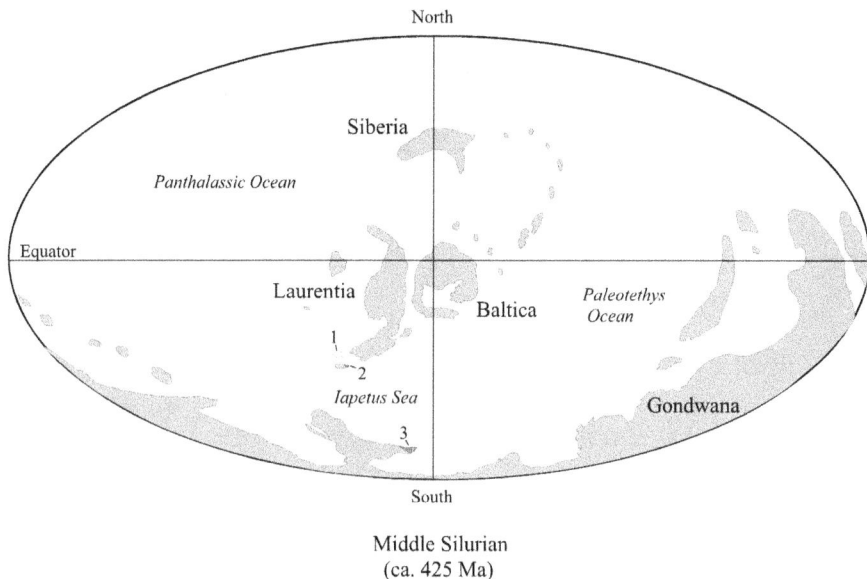

Middle Silurian
(ca. 425 Ma)

Figure 16 — Location of the three parts of Virginia (1-3) during the Middle Silurian Period. Back then, part 1 lay beneath the sea while parts 2 and 3 were on land. See Appendix 2 for the source of the base map of this picture.

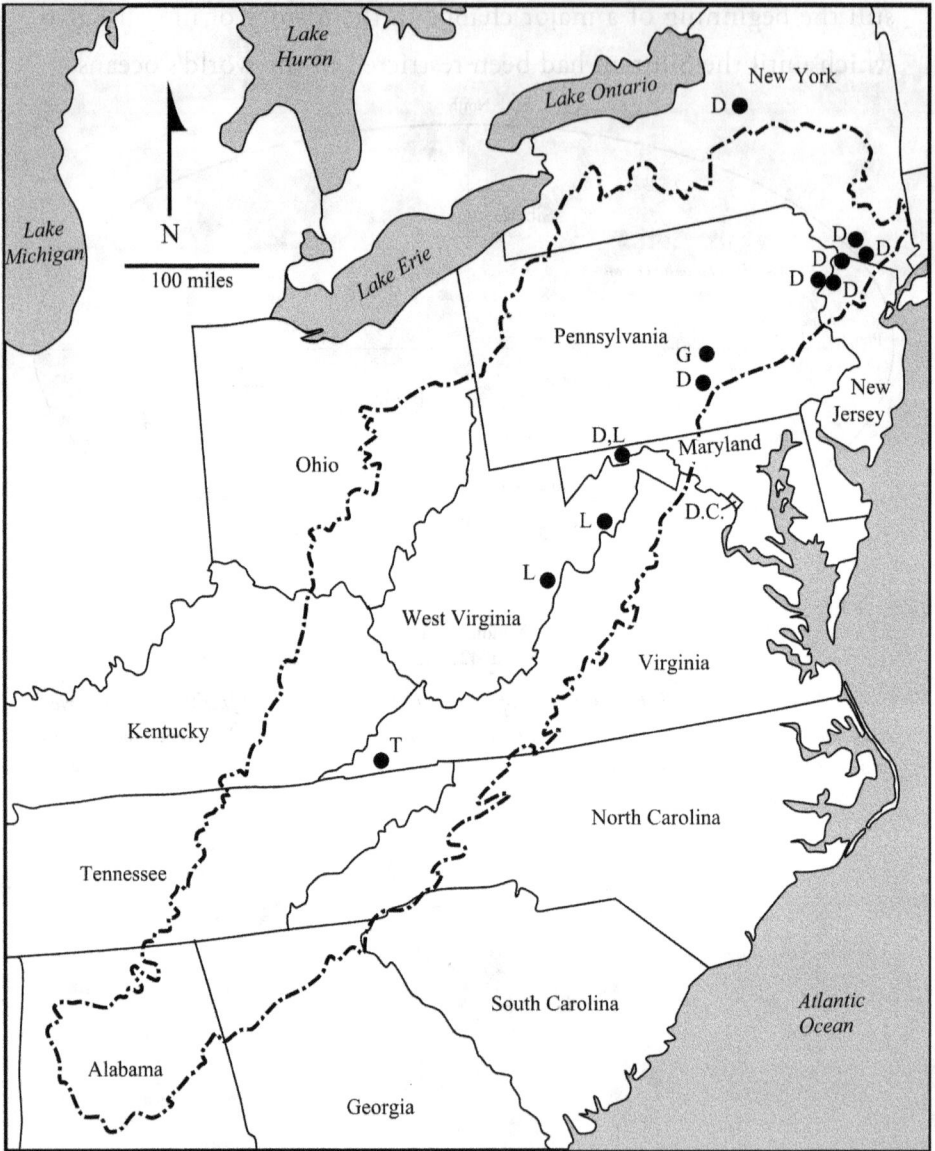

Figure 17 — Localities from which Silurian fishes have been described in and near the Appalachian region. D = localities described in Denison (1964), L = localities described in Leutze (1960), G = locality described in Giffin (1979), T = locality described in Weems and Grimsley (2022). See Appendix 2 for the source of this picture.

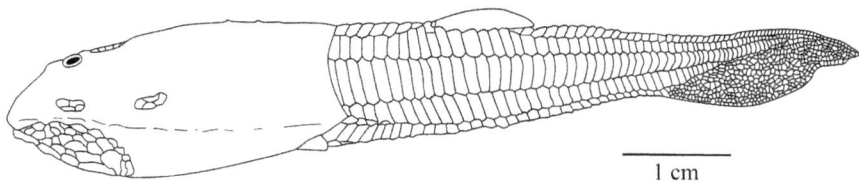

Tremataspis

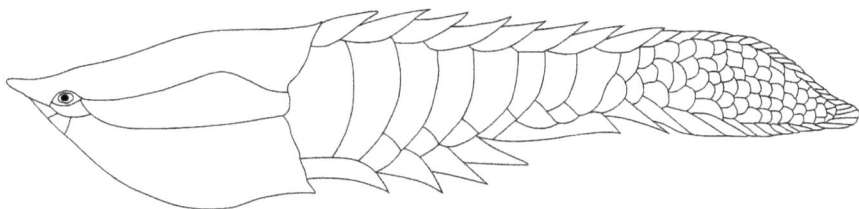

Poraspis

Thelodus

Figure 18 — Types of jawless fishes known from the Silurian Period in the Appalachian region. Tremataspis belonged to the class Osteostraci; Poraspis belonged to the class Heterostraci (as did the Appalachian genera Americaspis and Vernonaspis); Thelodus (and Logania) belonged to the shark-like class Thelodonti. Tremataspis is known from the Early Silurian; Americaspis, Vernonaspis, Thelodus, and Logania are all known from the Late Silurian. See Appendix 2 for the sources of this picture.

CHAPTER

7

THE DEVONIAN:

FISH BECOME ABUNDANT AND DIVERSE

With the onset of the Devonian Period, fishes in the Appala-chian region began to undergo a major interval of innova-tion and radiation (Figure 19). While this was going on, the number of co-existing conodont species dropped notably from sixty-four to forty-five types. While still numerous, the dominance of conodonts in the marine realm was clearly beginning to wane. At the same time, true fishes began to become much more common in the Appalachian region and increased in abundance upward from one species in the Early Devonian to thirty species by the Late Devonian.[26] Notably, these new species of fishes had lower jaws and also had fins along the sides of their bodies. These were major innovations in fish evolution, and they quickly became the dominant body plan to be found. Both of these innovations made Devonian fish much more successful and widespread throughout the oceans of the world than they had been previously during the Silurian Period.

At the beginning of the Devonian, fish were far more common in the shallow interior seaway than they were in the deeper waters of the offshore Appalachian seaway. Numerous fish have been described from the Early Devonian interior North American seaway region, but only one species of fish (*Machaeracanthus*) has been described so far from the Early Devonian marine deposits of the Virginia region.[27] In contrast, by the Late Devonian fishes had become much more abundant and numerous in the Appalachian Basin (Figures 20 and 21). Placoderm fishes appeared in the form of *Bothriolepis virginien-sis* in Virginia, *Glyptaspis eastmani* in Maryland and West Virginia, and *Bothriolepis nitida, Groenlandaspis pennsylvanica, Holonema rugo-sum, Phyllolepis rossimontina,* and *Turrisaspis elektor* in Pennsylvania (Table 1). Additionally, acanthodian fishes (*Gyracanthus*), shark-like chondrichthyan fishes (*Ageleodus*, *Ctenacanthus*), an actinoptery-gian (bony) fish (*Limnomis*), and sarcopterygian (lobe-finned) fishes (*Ganorhynchus*, *Holoptychius*, *Hyneria*, *Osteolepis*, and cf. *Sauripterus*) all made their first appearances in the Appalachian region within the state of Pennsylvania.[28] The explosion and spread of fish species in the Devonian seas was so dramatic that this period is often referred to as "the age of fishes."

In addition to the rapid spread of jawed fishes in the sea, some members of the sarcopterygian branch of fishes began to move into freshwater habitats and from there into terrestrial environments on the land (Figure 22). By this time in the Devonian, plants had also changed and adapted to spread into nearly all low-lying wet habitats near the sea. These plants were soon followed to land by arthropods, which gave rise to very large millipedes, centipedes, scorpions, spi-ders, and insects. Their spread into these environments brought forth an accompanying wave of carnivorous fish species, most notably those fishes that had begun to evolve into amphibians. As a result of these advances, before the end of the Devonian, two skeletal species of am-phibians are known to have been present in the western lowlands of Pennsylvania: *Hynerpeton bassetti* and *Densignathus rowei* (Figure 23). Footprints made by these or similar animals have so far not been discovered in Appalachia, but Devonian footprints are known from nonmarine Middle Devonian beds in Ireland and possibly also from nonmarine Upper Devonian beds in northern Scotland.[29] Therefore, it is likely that footprints will eventually be discovered in the Late Devonian rocks of Appalachia. The appearance of amphibians in the Devonian marks the beginning of a major wave of evolutionary ad-vancement in which amphibians that were descended from fish began to move out of the water and onto the now-vegetated lands.

Near the end of the Devonian, the Appalachian seaway ceased to be dominated by marine beds and instead began to be covered by continental red beds. This transition from dominantly marine to dominantly terrestrial beds continued throughout the later history of what had been the Appalachian seaway, and includes the strata deposited there during the following Mississippian, Pennsylvanian, and Early Permian periods. This change in depositional history seems largely to have been the result of a strong drop in world sea level, which was caused by the beginning of a long interval of major ice ages in the southern hemisphere continent of Gondwana (Figure 16). This event began in the Late Devonian with the development of large glacial ice sheets in what today are Brazil and Bolivia. From there, more ice sheets began to appear in what are now Antarctica, Australia, southern Africa, India, and Arabia, until all the land masses of the southern hemisphere came to lie beneath large glacial ice sheets. This climatic pattern continued to be the global norm until at least the middle of the Permian Period.[30]

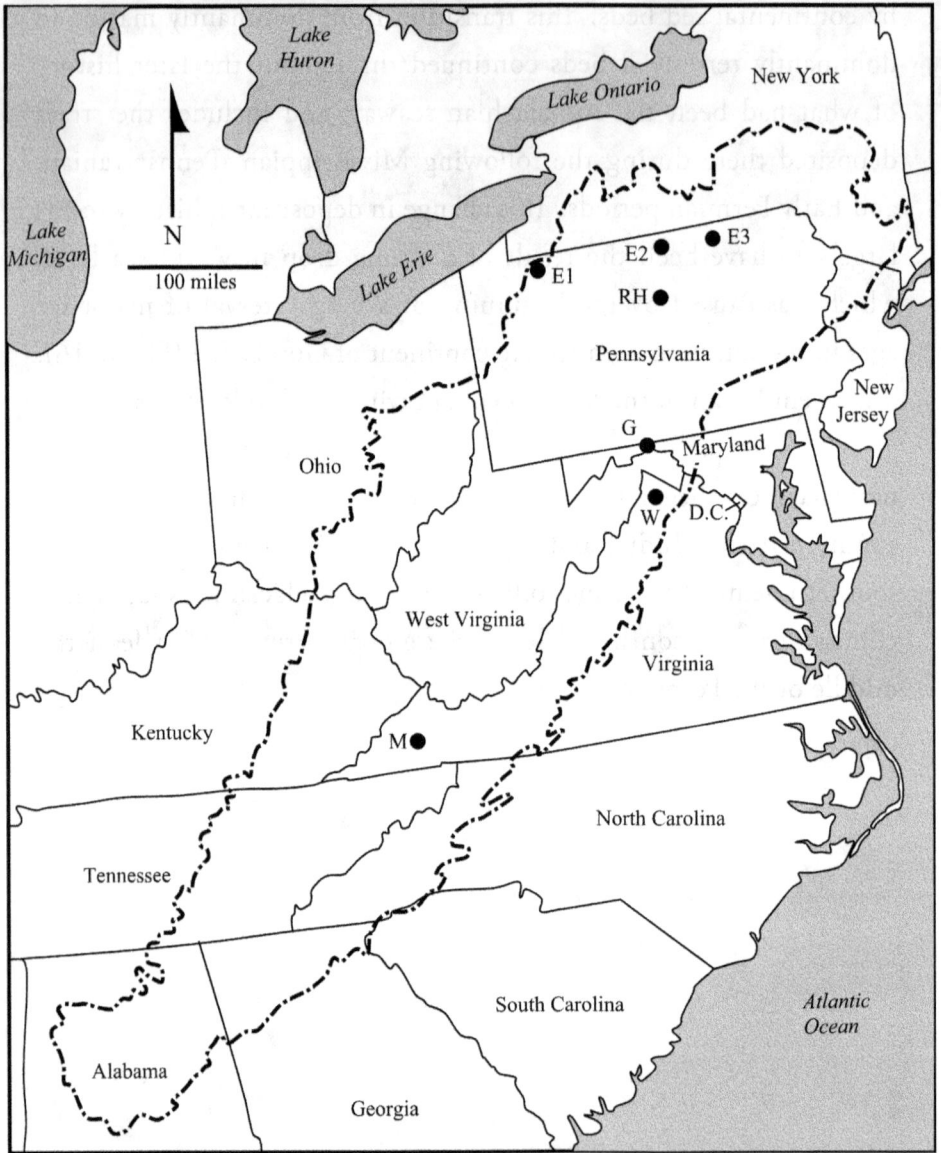

Figure 19 — Major locations where Devonian vertebrate fossils have been found in the Appalachian region. E = Eastman (1908) sites; G = Glyptaspis site (Swartz, 1913); M = Machaeracanthus site (Butts, 1940), RH = Red Hill site (Daeschler and Cressler, 2011) which has yielded numerous genera of fossil fish and amphibians; W = Weems et al. (1981) Bothriolepis virginiensis site and Butts and Edmundson (1966) Eczematolepis (=Acantholepis) site. See Appendix 2 for the source of this picture.

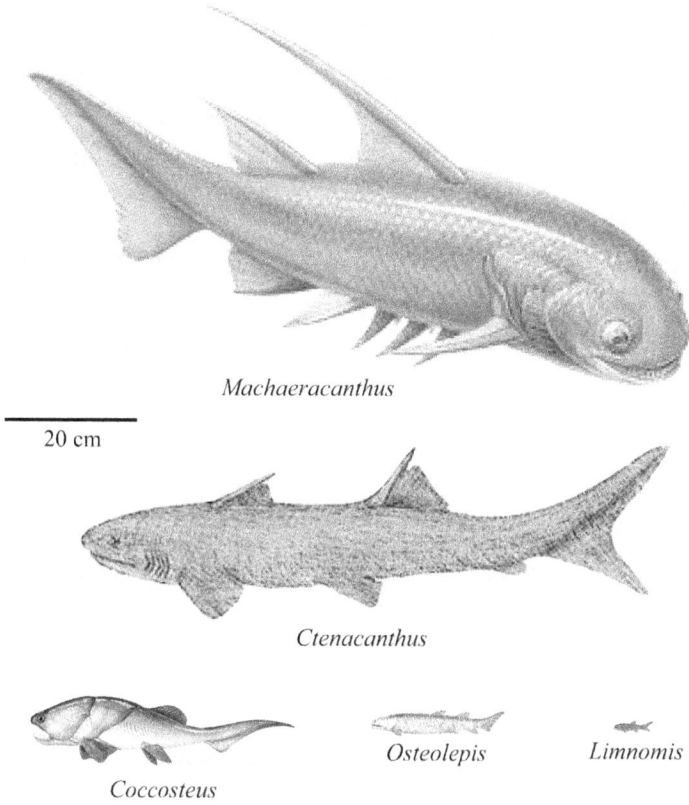

Machaeracanthus

20 cm

Ctenacanthus

Coccosteus

Osteolepis

Limnomis

Figure 20 — Typical Devonian fossil marine fishes known from the Appalachian region. See Appendix 2 for the sources of this picture. Machaeracanthus *is known from Lower Devonian beds, while the other fishes have been found in Upper Devonian strata.*

Figure 21 — Bothriolepis virginiensis, *a Late Devonian fossil marine fish from Virginia. See Appendix 2 for the source of this picture.*

Figure 22 — A Middle Devonian landscape, showing typical plants in a marsh setting. Sites like this were where crossopterygian fishes first began to move out of the water and develop into the earliest amphibians. See Appendix 2 for the source of this picture.

Hynerpeton

Pederpes

0.5 meter

Figure 23 — Two kinds of Upper Devonian amphibians known from the Red Hill site in Pennsylvania. Hynerpeton *is directly identified from this site. A second kind of amphibian, only known to be a member of the family Whatcheeridae, is represented here by* Pederpes *from Scotland, which is a member of this same family. See Appendix 2 for the sources of this picture.*

CHAPTER

8

THE EARLY CARBONIFEROUS (MISSISSIPPIAN): VERTEBRATES MOVE ONTO LAND

In Europe and many other parts of the world, the second to last period of time in the Paleozoic Era is known as the Carboniferous. In America, however, this interval of time has long been recognized as having two rather different natures, and for this reason, it was divided into two different intervals which were named the Mississippian Period and the Pennsylvanian Period. Because the Carboniferous is one of the longer periods in the geologic column, and because this is America, we use the American time system here and divide the Carboniferous into two separate intervals.

The record of fossil vertebrates in the Appalachian region is much better known in the Mississippian Period than it was in the Devonian (Figure 24). The known Mississippian fossils include both skeletal remains and, in the case of amphibians and reptiles, footprints. The footprints tell us that some of the amphibians and reptiles that were present here are not known from bony remains that have been found within this interval of time in the Appalachian region. Clearly, there is much that remains to be learned about the animals that lived in the Virginia region during this interval of time.

By the Mississippian Period, the sea between Laurasia-Baltica and Gondwana had begun to narrow considerably (Figure 25). Because of the ongoing glacial age in the southern hemisphere, which globally lowered sea levels around the world, the seas in the eastern Appalachian region became very shallow or disappeared entirely. As a result, the Appalachian region was only covered by a shallow sea for a short span of time during the Middle Mississippian Period. During the rest of this time, the Appalachian region was above sea level and was only home to freshwater and onshore animals and plants.

The Mississippian Period began at the end of a major extinction

event among the fishes and the conodont animals. While this extinction event at the end of the Devonian was a major one, its ultimate
cause remains a mystery. The placoderms, which had done very well
during the Devonian, abruptly became completely extinct. In the
Appalachian basin, the acanthodians were reduced to a single genus,
and the conodont animals were reduced to eleven named types that
probably represented only two biological species. Only two species of
Mississippian actinopterygian (bony) fishes have been discovered and
named, and only one species of sarcopterygian (lobe-finned) fish has
been reported (Figure 26). The only group to show a major increase in
abundance and number of new species at this time was the chondrichthyan (shark-like) fishes, which increased in diversity from two to elev-
en species. The massive decline in the abundance of the conodont
animals at the same time that the shark-like animals were expanding
in diversity suggests that predation by the shark-like animals may have
been a considerable reason for the decline of the conodont creatures,
which probably lived in the open ocean waters that the sharks were
coming to dominate.

The other feature that characterizes the Mississippian Period is the
diversification of the amphibians during this interval of time. In the
Appalachian region, the two skeletal amphibian species known from
the Late Devonian were equal in relative abundance to the two skeletal
amphibian species known from the Mississippian Period (Figure 27).
Even so, fossil footprints from Mississippian rocks (Figure 28), especially in Pennsylvania, indicate that at least four kinds of amphibians
must have been present.[31] Therefore, amphibians were likely at least
twice as diverse in the Mississippian as they had been in the Devonian. Additionally, two other types of fossil footprints, including one
(*Dromopus*) found in western Virginia, indicate that there were also
two kinds of early reptiles within this terrestrial realm. This can be
surmised because reptiles have claws on their toes, while claws are not
present on the feet of amphibians.[32]

The diversification of amphibians and reptiles probably occurred because of the increasing abundance of tree-like and brushy plants on the surface of the land. Notably, one of the oldest coal deposits in the world is found in the Upper Mississippian strata near Blacksburg, Virginia.[33] The spread of these forests provided shade for these early land vertebrates and cover from predators as they began to move more and more out of the aquatic realm and take up life permanently on the land.

Figure 24 — Localities from which Mississippian vertebrates have been described in the Appalachian region. A = Arkle et al. (1979), B = Branson (1910), F = Fillmore et al. (2012), M = Mickle (2018), and W = Weems and Windolph (1986). See Appendix 2 for the source of this picture.

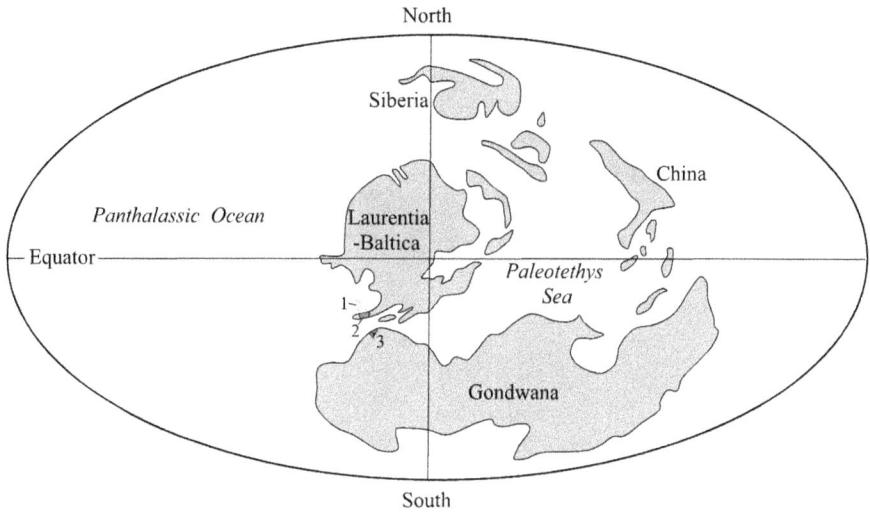

Early Carboniferous (Mississippian)
(ca. 425 Ma)

Figure 25 — Location of the three parts of Virginia (1-3) during the Early Carboniferous (Mississippian) Period. Back then, all parts of Virginia lay slightly south of the equator and were oriented along a northwest-southeast trend. See Appendix 2 for the source of this picture.

Gyracanthus

10 cm

Stethacanthus

Hadronector

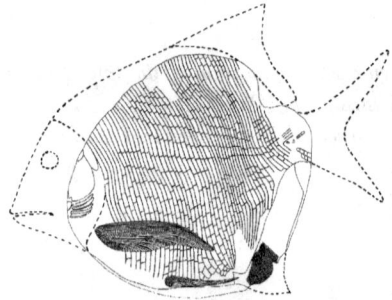

Tanypterichthys

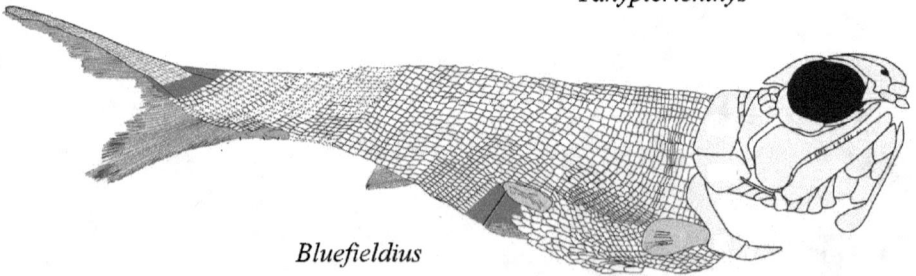

Bluefieldius

Figure 26 — Mississippian fishes known from the Appalachian region. Gyracanthus *is an acanthodian fish.* Stethacanthus *is a shark-like chondrichthyan fish closely related to genera known from the Appalachian region.* Tanypterichthys *and* Bluefieldius *are bony fishes.* Hadronector *is a close relative of the Appalachian lobe-finned fish* Tranodis. *See Appendix 2 for the sources of this picture.*

Greererpeton

1 meter

Proterogyrinus

Figure 27 — Although footprints indicate four kinds of amphibians were present in the Mississippian Appalachian region, we have skeletal remains of only two of them. Greererpeton *likely made the tracks known as* Batrachichnus; Proterogyrinus *likely made the tracks known as* Pseudobradypus. *See Appendix 2 for the sources of this picture.*

Amphibians   Reptiles

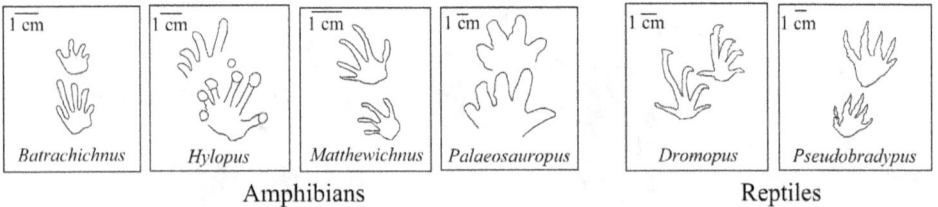

Figure 28 — Footprints of amphibians and reptiles found in the Mississippian rocks of the Appalachian region. The top picture shows a temnospondyl amphibian that was likely close to the trackmaker of the Palaeosauropus *footprints. All of these Mississippian animals produced similarly sprawling track patterns. See Appendix 2 for the sources of this picture.*

CHAPTER

9

THE LATE CARBONIFEROUS (PENNSYLVANIAN):

THE COAL AGE

As was the case during the preceding Mississippian Period, the main region that contains a record of the Pennsylvanian Period is located in the far western part of Virginia and nearby areas in Appalachia (Figure 29). At that time, areas to the east of this region were undergoing profound erosion due to the continuing rise of the ancestral Appalachian Mountains. A large part of the materials eroded from this rising mountain region was transported westward along rivers toward the western part of what had been the Appalachian seaway, creating a large area of river bottoms and swamp lands that were accumulating thick deposits of coal. These deposits include most of the major layers of Appalachian coal that have been mined in this region over the last two centuries.

The Pennsylvanian Period continued the general trends of vertebrate evolution that had begun in the Mississippian Period. Conodonts and acanthodian fishes managed to stabilize and maintain their strongly reduced numbers and reduced diversity in the seas. In contrast, sharks, bony fish, and lobe-finned fish continued to increase in numbers and diversity, expanding their hold on the marine realm (Figure 30). During this same interval of time, the bony fish and lobe-finned fish continued to expand their diversity in freshwater rivers and lakes. In these latter environments, amphibians also continued to diversify and filled an increasing number of niches in the coal forest environments that were steadily becoming more widespread throughout the world at this time (Figure 31). Reptiles also continued to diversify, but in addition, new kinds of creatures appeared called "mammal-like reptiles" that were slowly beginning to develop into true mammals (Figure 32). The mammal-like reptiles represented only a small component of the coal swamp faunas of that time, but they would later come to be far more abundant.

During the Pennsylvanian, coal swamps spread widely across the equatorial regions of the world and huge volumes of coal were deposited.[34] These swamps captured large volumes of carbon, removing it from the Pennsylvanian atmosphere and making the climate much richer in oxygen than it had been at any time earlier or since. One result of this seems to have been the appearance of very large land-dwelling arthropods that came to grow to a gigantic size relative to the ones that we see today (Figure 33). Millipedes left tracks indicating that they included individuals that were up to two meters (six feet) long.[35] Fossils of dragonflies have been found that have wingspans that are more than 70 centimeters (two feet) across.[36] The presence of these and other very large arthropods suggests that they were able to get so big during that time because of the very high oxygen levels that the coal-swamp plants were then producing. Unfortunately for these animals, this global abundance of oxygen did not last long beyond the end of the Pennsylvanian.

Another major event in world history during the Pennsylvanian was the continuing cyclic expansion and contraction of glaciers across the polar regions of the world.[37] This caused a periodic fluctuation in global sea levels as these glaciers waxed and then waned over time due to cyclic climate fluctuations. The episodic growth and melting of these glaciers created a cyclic pattern of deposition in the Appalachian region that has left us with repetitive layers of thick coal beds that are separated by layers of shale, sandstone, and rare interbedded marine limestones.

In Virginia and other Appalachian regions, the organic matter left behind as coal also strongly acidified the groundwater that was present. This in turn tended to dissolve most of the calcium and phosphorus deposits that were forming within them. As a result of this, remains of vertebrates are generally rare, though exceptionally they do occur. Occasionally, footprints of amphibians show up (Figure 33) which tell us that these animals were still fairly common in the Appalachian region even though their bony remains tend to be absent. Overall, thirty kinds of amphibians have been recognized and described from the Pennsylvanian-age beds preserved in the Appalachian region, as well as two kinds of reptiles and two kinds of mammal-like reptiles. Clearly, amphibians were still the most abundant and dominant life forms in the coal swamps of the Appalachian world. Even so, the appearance of reptiles and mammal-like reptiles hinted that significant changes were soon to follow in the faunal composition of the late Paleozoic continents.

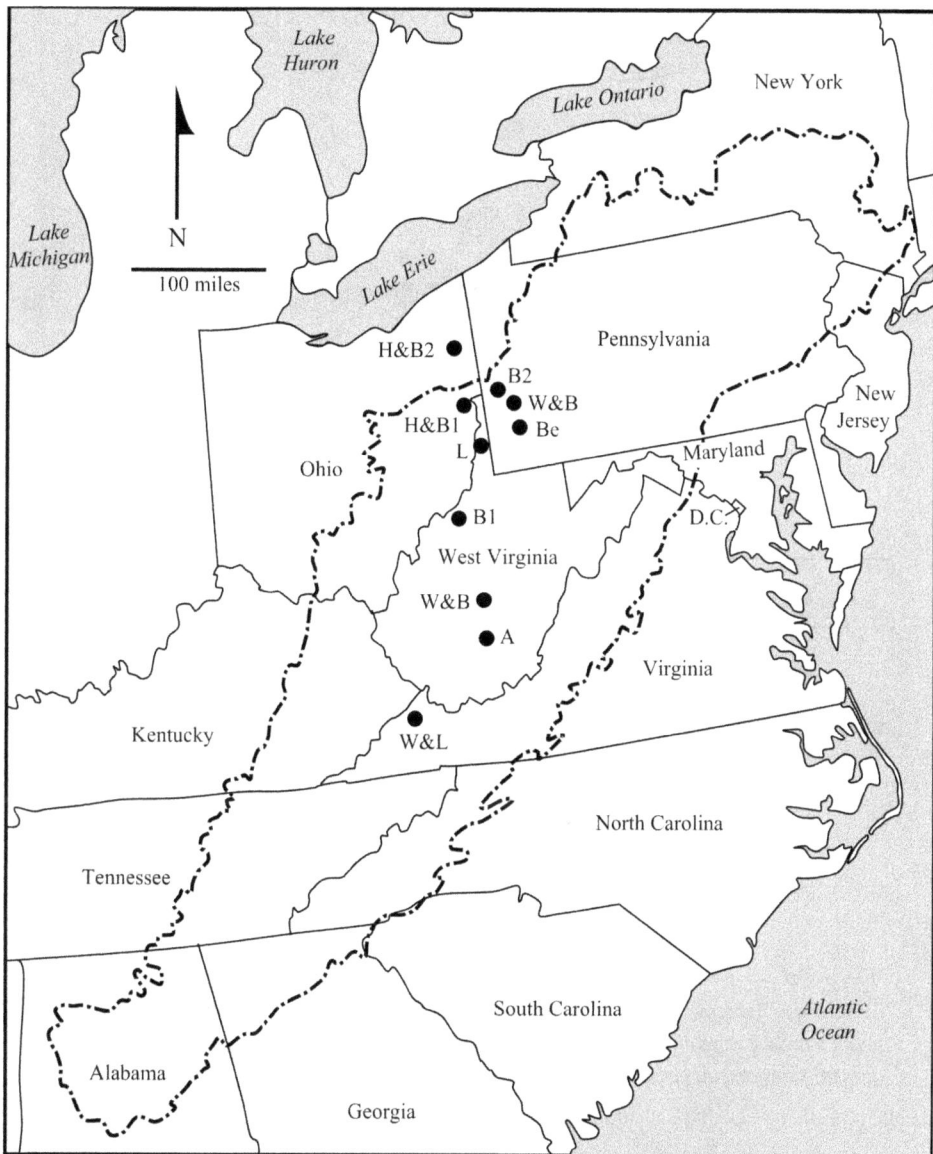

Figure 29 — Localities from which Pennsylvanian vertebrates have been described in the Appalachian and nearby regions. A = Arkle et al. (1979), B1 = Baird (1957), B2 = Baird (1978), Be = Berman et al. (2010), L = Lund (1972a), H and B1 = Hook and Baird (1988), H and B2 = Hook and Baird (1994), W and B = Werneburg and Berman (2012), W and L = Weems and Lucas (2021). See Appendix 2 for the source of this picture.

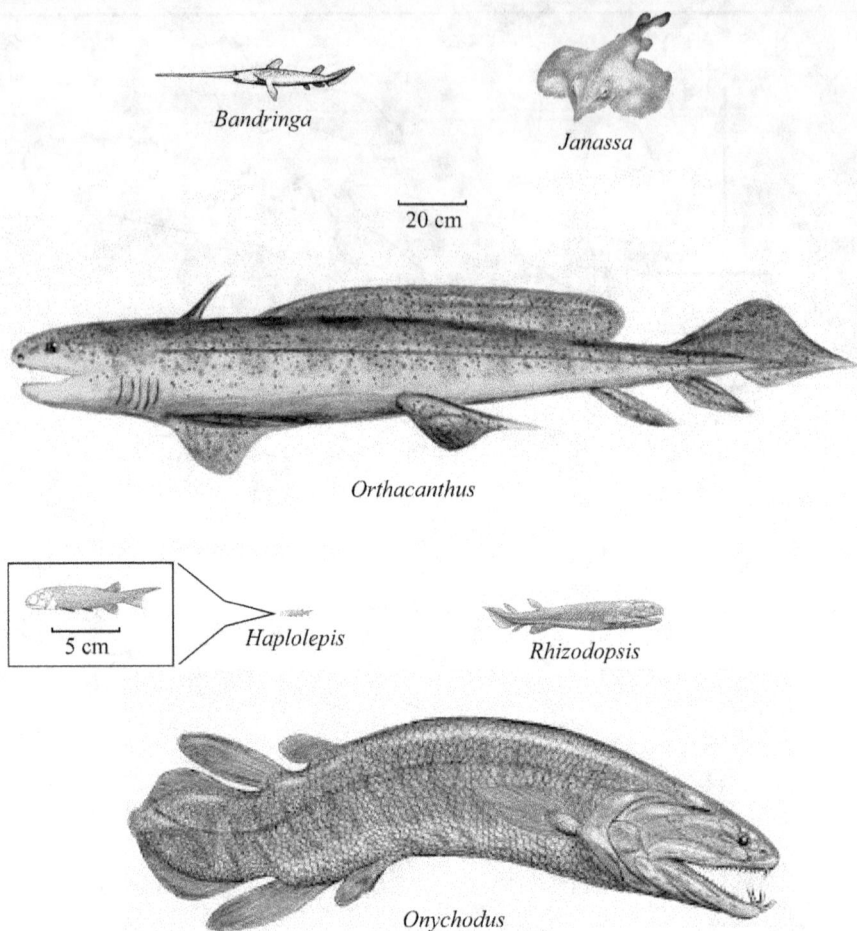

Figure 30 — Fossil fish were widespread in the Pennsylvanian seas and swamplands. Bandringa, Janassa, and Orthacanthus *were Pennsylvanian shark relatives;* Haplolepis *was an actinopterygian (bony) fish;* Rhizodopsis *and* Onychodus *were sarcopterygian (lobe-finned) fishes. See Appendix 2 for the sources of this picture.*

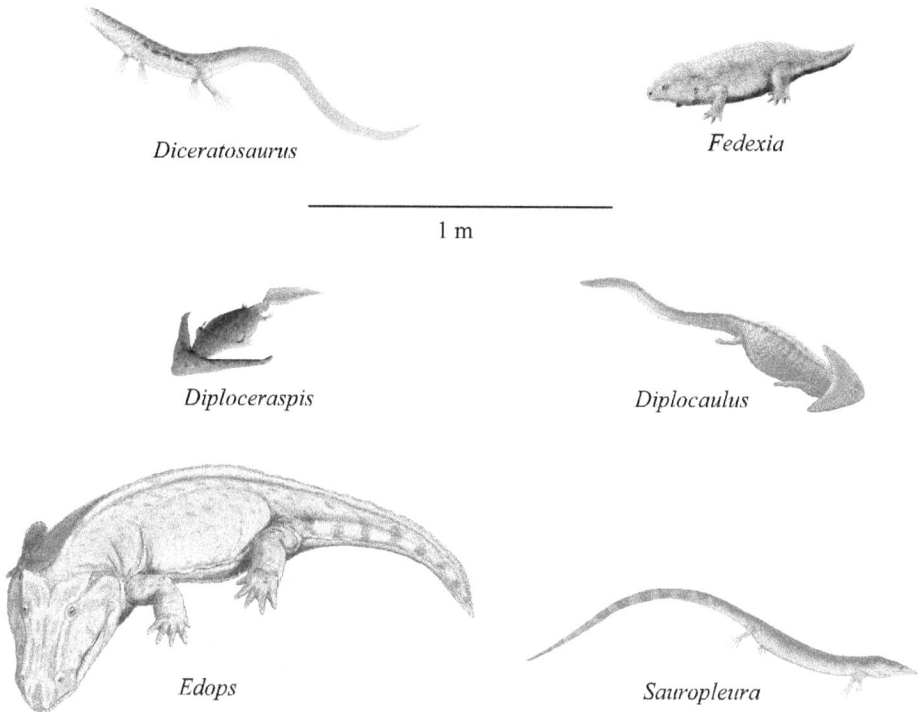

Diceratosaurus

Fedexia

1 m

Diploceraspis

Diplocaulus

Edops

Sauropleura

Figure 31 — The Pennsylvanian was the age in which amphibians had their greatest abundance and diversity. See Appendix 2 for the sources of this picture.

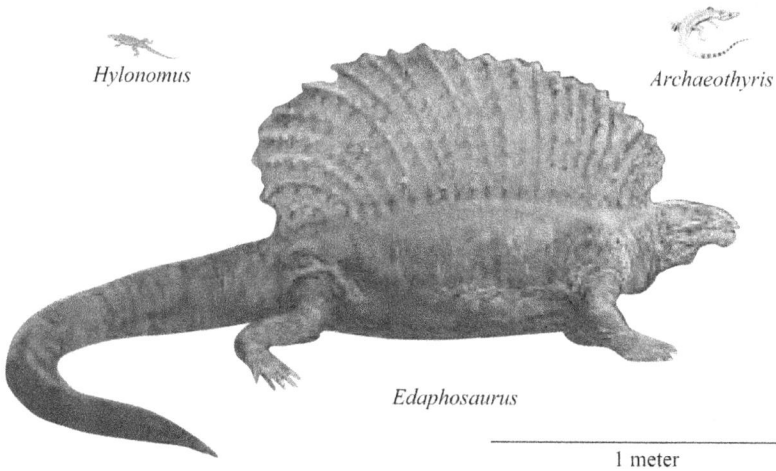

Hylonomus

Archaeothyris

Edaphosaurus

1 meter

*Figure 32 — One of the earliest known reptiles (*Hylonomus*) and two mammal-like reptiles (*Archaeothyris *and* Edaphosaurus*) made their first known appearances in the eastern American coal swamps during the Pennsylvanian Period. See Appendix 2 for the sources of this picture.*

Figure 33 — (Top) In the Pennsylvanian Appalachian region, forests were dense and produced huge deposits of peat in swampy regions. This peat became buried and over time compressed into coal. (Bottom) Some arthropods grew to extremely large size in the coal swamps of the Pennsylvanian, such as the millipede Arthropleura *and the dragonfly* Meganeura. *See Appendix 2 for the sources of this picture.*

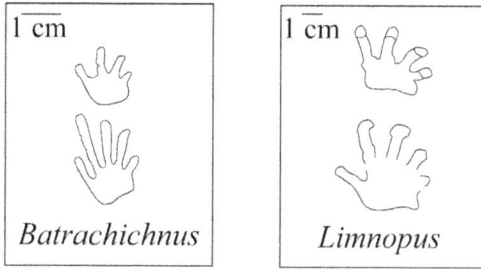

Amphibians

Figure 34 — Footprints of amphibians known from the Pennsylvanian rocks of western Virginia. See Appendix 2 for the source of this picture.

CHAPTER

10

THE PERMIAN:
THE FINAL ACT OF THE PALEOZOIC ERA

During the early part of the Permian Period, which ended the Paleozoic Era, the Appalachian Mountains continued to rise and the area affected by their uplift continued to extend ever farther westward. As a result of this, no areas underlain by Early Permian rocks still exist within the modern state of Virginia, though deposits of that age do persist in nearby Maryland, Ohio, Pennsylvania, and West Virginia (Figure 35). As in the Pennsylvanian Period before it, Permian material eroding from the rising Appalachian mountain chain washed westward across what was left of the Appalachian seaway, adding river bottom and swamp land deposits to this region and continuing to deposit coal for a while as had occurred during the Pennsylvanian Period. Later in the Permian, as the Appalachian seaway became totally infilled, the eastern and western parts of Virginia became geographically continuous as they are today, and the Virginia region became greatly uplifted and isolated within the continent of Pangea far from any nearby sea waters (Figure 36).

The Permian Period brought to a close the remarkable series of advances that vertebrates had achieved since their earliest appearance in the Cambrian Period. Nowhere in the Appalachian region are vertebrate-bearing rocks of Middle or Late Permian age definitely known to exist, so events toward the end of the Permian Period are of necessity inferred from what is preserved in other regions of the world today. These distant outcrops tell us that the end of the Permian was a very harsh time for the animals that lived in the world at that time.

Although acanthodian fish apparently became extinct in the Appalachian region during the Pennsylvanian Period, a single species (*Acanthodes marshi*) re-established the presence of this group here in the subsequent marine realm of the Early Permian (Figure 37). This was not to last, however, for acanthodian fishes globally became extinct by the end of the Permian. Similarly, at least one conodont animal survived into the Permian and a successor species of this group continued to exist elsewhere in the world until the end of the subsequent Triassic Period, early in the Mesozoic Era. With this one exception, all of the more primitive Paleozoic fishes entirely ceased to exist by the end of the Permian. All of the other types of fish that survived into the Mesozoic Era were sharks, bony fish, and lobe-finned fish.

The Permian Period was marked by two of the greatest extinction events that are known to have occurred among the terrestrial and marine animals on planet Earth. At the end of the Middle Permian, there was a very large extinction event among the terrestrial animal species on Earth, during which 70% of known terrestrial vertebrate species died out. Although the fossil insect record of the Permian remains poorly known, what has been found seems to show that their greatest known decline in diversity also occurred at about this time. [38] The second known extinction event at the end of the Permian was so profound that it saw the disappearance of 81% of all known marine species, including the characteristically Paleozoic trilobites.

It was long assumed that the terrestrial and marine extinction events of the Permian occurred at the same time. However, recent advances in age dating in the later Permian have revealed that most of the terrestrial extinctions occurred at the boundary between the Middle Permian and the Late Permian, which was somewhat earlier than the main marine extinction event. [39] The mid-Permian terrestrial extinction event has been linked to the eruption of the Emeishan Traps, a vast volcanic flow deposit formed during a large igneous event in southwestern China about 260 Ma. [40] Similarly, the end-Permian marine extinction event has been linked to the eruption of the Siberian Traps, which formed an even larger outpouring of basaltic lavas in the region of modern Siberia around 252 Ma. [41] These two major volcanic events greatly disrupted the stability of the world's land and atmospheric environments and subsequently led to a major disruption in the stability of the world's oceans. When this combined interval of two great volcanic eruptions was finally finished, the fauna of the world had been greatly reduced and what followed was a strikingly different array of animals that came to characterize the Mesozoic Era. These new groups notably included the crocodilians, dinosaurs, and birds.

Unfortunately, the fossil record in the Appalachian region does not include a record of these major geological and biological catastrophes. Most of the Early Permian story is preserved in the rocks of Appalachia, but the story of the remainder of that period was either never laid down here or more likely later eroded away. In eastern Virginia, pollen of either Middle or Late Permian plants has been discovered in Triassic rocks near Ashland, Virginia.[42] This reworked relict pollen hints that it is possible that fossiliferous rocks of Middle or Late Permian age may someday be discovered in this region. For now, though, all we have as a record of the later Permian Period in the eastern United States are these few genera of reworked Middle to Late Permian pollen grains.

What we do find even in the Early Permian rocks of the Appalachian region is evidence of a declining diversity of bony fish. Although bony fishes would later come to dominate the fish faunas of the world, in the late Paleozoic they seem mostly to have simply held on and awaited better opportunities that would come in the subsequent Mesozoic and Cenozoic Eras. Sharks also declined due to the end-Permian extinction events, but they were able to re-diversify quickly at the beginning of the Mesozoic Era and became abundant again in the marine realms of the world up to the present day. In contrast to both of these groups, the lobe-finned fishes that had dominated the Early Permian aquatic environments underwent a strong decline both in their marine and nonmarine habitats. Today, we only know of two species of this once-diverse group of fish that have survived into modern times and only in the depths of the Indian Ocean. [43] To their great credit, lobe-finned fish succeeded in siring the entire array of land animals (amphibians, reptiles, birds, and mammals) that arose from them. But, despite their familial successes, lobe-finned fishes in their original form declined after the end of the Paleozoic and today are almost extinct.

During the Early Permian, amphibians underwent a slight decline in abundance and diversity, even though they were still the single most diverse group of vertebrates then living in the terrestrial world (Figure 38). Reptiles also survived, but it was the mammal-like reptiles of that time that showed the greatest increase in their abundance and diversity (Figure 39). During the Early Permian interval, all of these groups left footprints recording their activity in the Appalachian region (Figure 40). This pattern continued beyond the end of the known Appalachian story and correlates with the last of the coal-forming age, which probably persisted for a while in the Appalachian region during the remainder of the Early Permian.[44]

After the Early Permian, the world climate became drier as Pangea fully united with Laurasia and both regions together drifted northward until the central region of this continent became equatorial desert lands. As a result of these changes, amphibians declined markedly and reptiles were simply able to maintain a steady existence. In contrast, mammal-like reptiles adapted and evolved to fit into this new world environment throughout the middle part of the Permian. Little is known of Middle Permian life in America, but a great many kinds of fossil-bearing deposits of this age have been found in the Ural Mountains region of Russia and in South Africa. These deposits give us our best available view of what was happening to land life in the middle part of the Permian Period (Figure 41).[45]

Interestingly, the mammal-like reptiles continued to adapt and spread across the Middle Permian world right up until the beginning of the Late Permian. Then, they abruptly underwent a massive decline.[46] Why they largely died off then, with reptiles and parareptiles becoming dominant in the latest Permian beds (Figure 42) and then reptiles surviving onward into the beginning of the Mesozoic Era, appears to be strongly linked to the massive Emeishan volcanic event in China. This disruption of the world's atmosphere by a major volcanic episode, along with the subsequent Late Permian Siberian Trap volcanic event, vastly destabilized the world's environment and resulted in a massive global warming and extinction event that brought the Paleozoic world to an end. While it is unfortunate that we have no direct evidence of these events in the Virginia region, the greater global record clearly indicates that we were not spared from the effects of this major turnover in life at the end of the Paleozoic. When the Virginian geologic record resumed during the Triassic Period of the subsequent Mesozoic Era, the world of the Paleozoic had largely been erased and a new world pattern had begun to take shape.

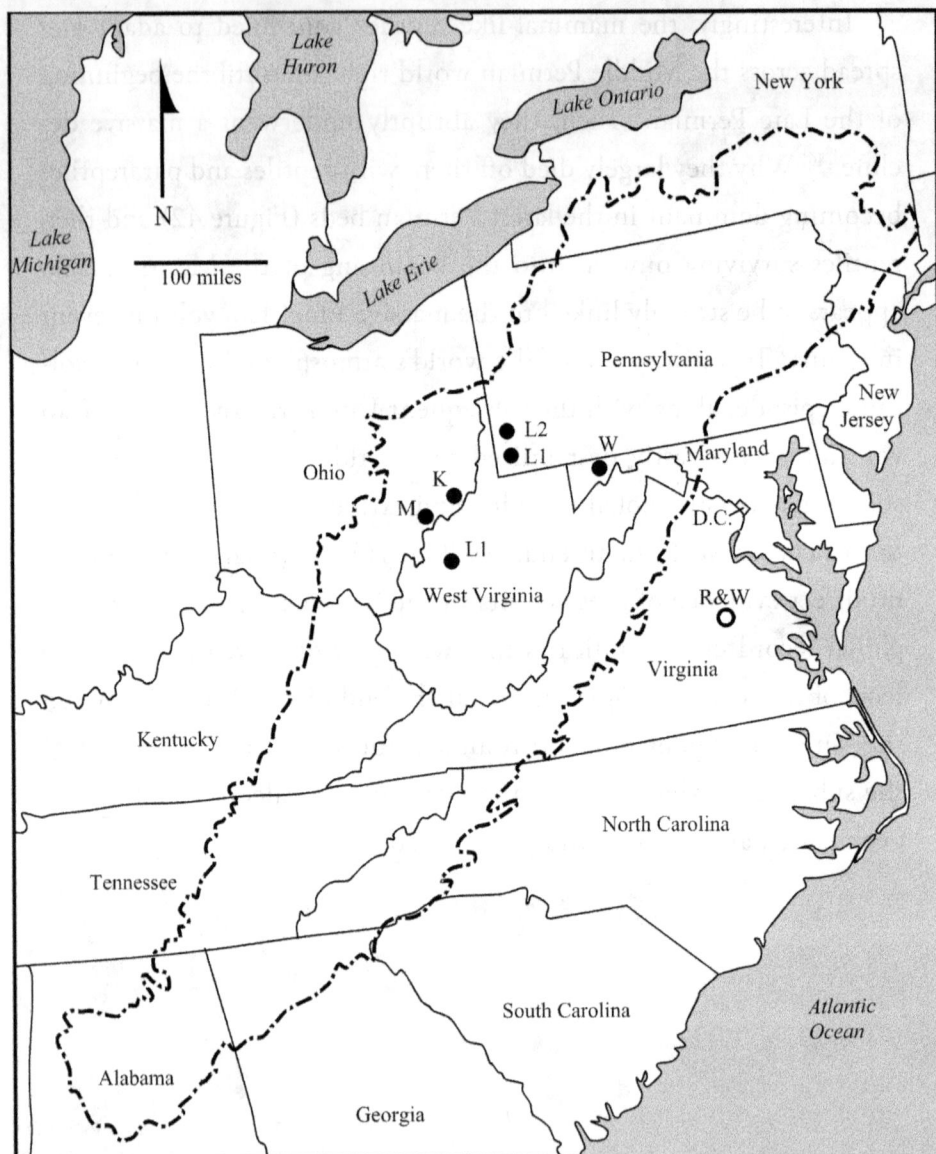

Figure 35 — Black circles — Localities from which Early Permian vertebrate remains have been discovered in the Appalachian region; White circle — Locality where reworked Middle or Late Permian pollen has been recovered in Virginia; L1 = Lucas et al. (2016); L2 = Lund (1972b); M = Martin (1998); K = Kissel (2010); R and W = Robbins and Weems (1988); W = Weems (2022). See Appendix 2 for the source of this picture.

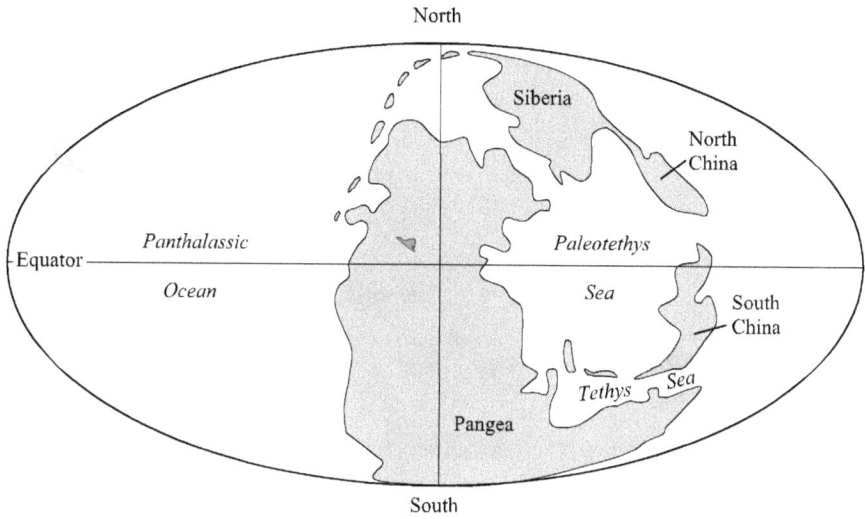

North

Siberia

North
China

Panthalassic

Equator

Ocean

Paleotethys

Sea

South
China

Tethys Sea

Pangea

South

Late Permian
(ca. 255 Ma)

Figure 36 — By the Late Permian Period, Virginia had become united into the single land area that we know today (dark gray). Also by this time, it had drifted northward to a location slightly above the equator in the tropical portion of the world. See Appendix 2 for the source of this picture.

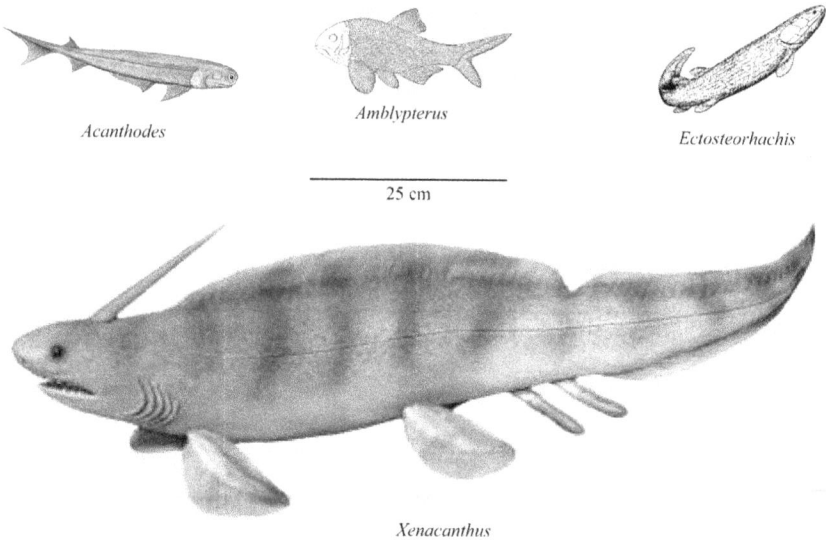

Acanthodes

Amblypterus

Ectosteorhachis

25 cm

Xenacanthus

Figure 37 — Early Permian fishes were similar to those that preceded them during the Pennsylvanian Period. Acanthodes *was one of the last of the acanthodian fishes.* Amblypterus *was an actinopterygian (bony) fish and* Ectosteorhachis *was a sarcopterygian (lobe-finned) fish. Xenacanthus was a very large shark-like chondrichthyan fish. See Appendix 2 for the sources of this picture.*

Diploceraspis

Megamolgophis

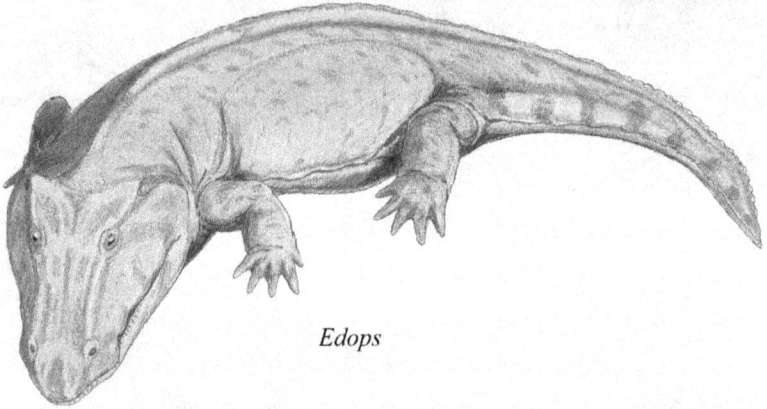

Edops

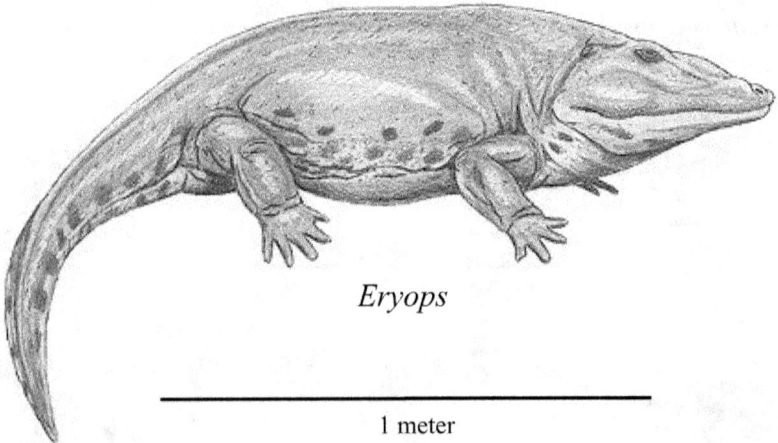

Eryops

1 meter

Protorothyris

Figure 38 — Although common at the beginning of the Permian, the number and variety of amphibians greatly declined as the world's climate became much drier during the later Permian. Reptiles, such as Protorothyris, *persisted but remained small. See Appendix 2 for the sources of this picture.*

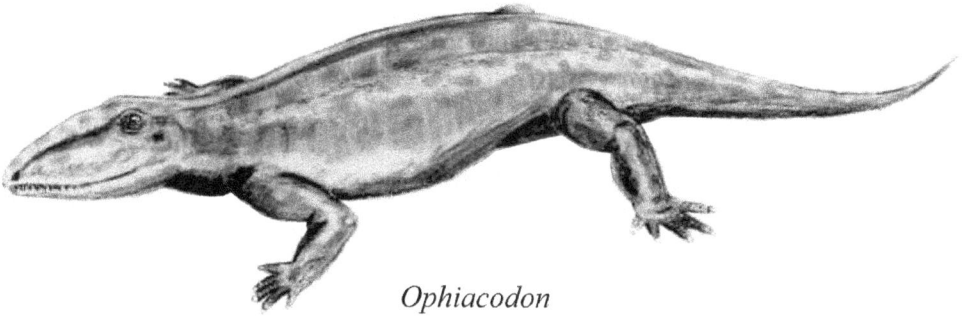

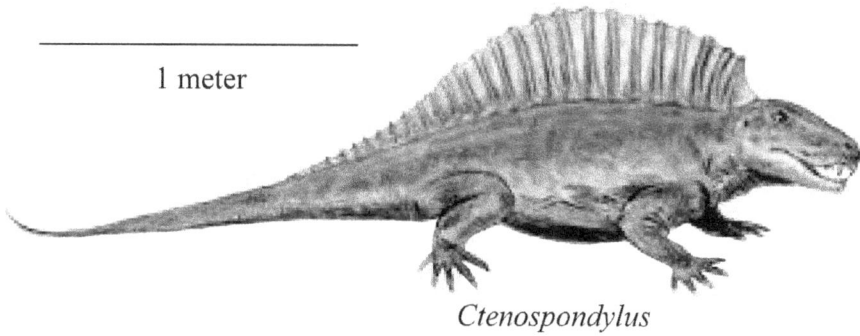

Figure 39 — Mammal-like reptiles continued to diversify in the Early Permian and became both larger and more abundant than they had been in the Pennsylvanian. See Appendix 2 for the sources of this picture.

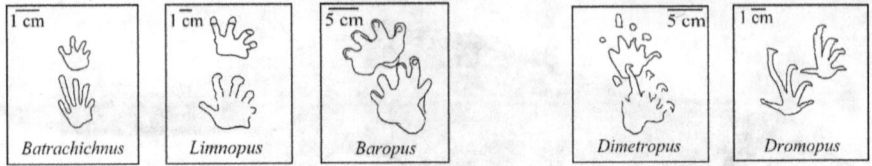

Amphibians

Reptiles

Figure 40 — Permian footprints are known from the Dunkard Group in Maryland, Pennsylvania, and West Virginia. See Appendix 2 for the source of this picture.

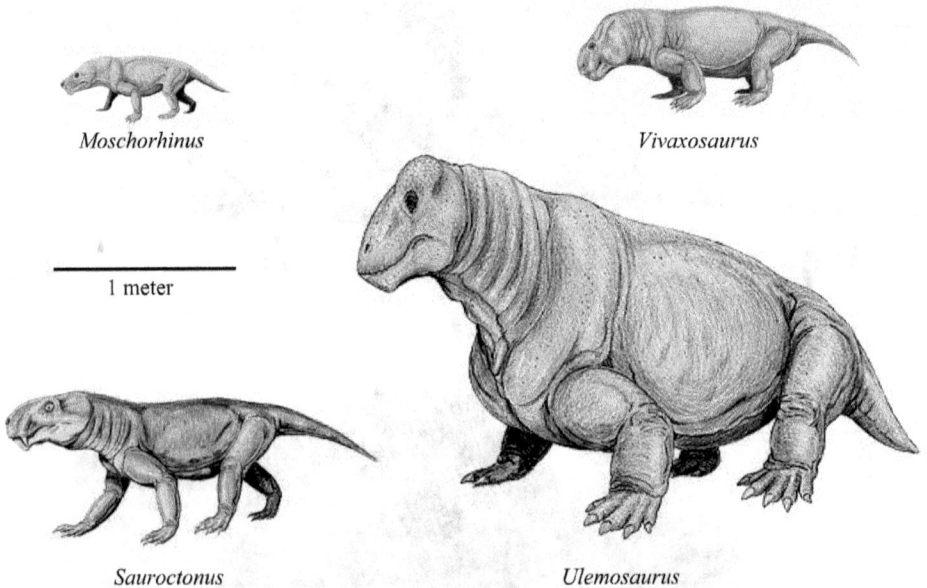

Figure 41 — A Middle Permian record is unknown in North America, but Middle Permian deposits in Russia and South Africa include fossils of animals that show the mammal-like reptiles continued to be dominant in the terrestrial environments of the world. Presumably, these or similar animals were also dominant in Virginia and nearby regions. Most of these mammal-like reptiles became extinct by the end of the Middle Permian. See Appendix 2 for the sources of this picture.

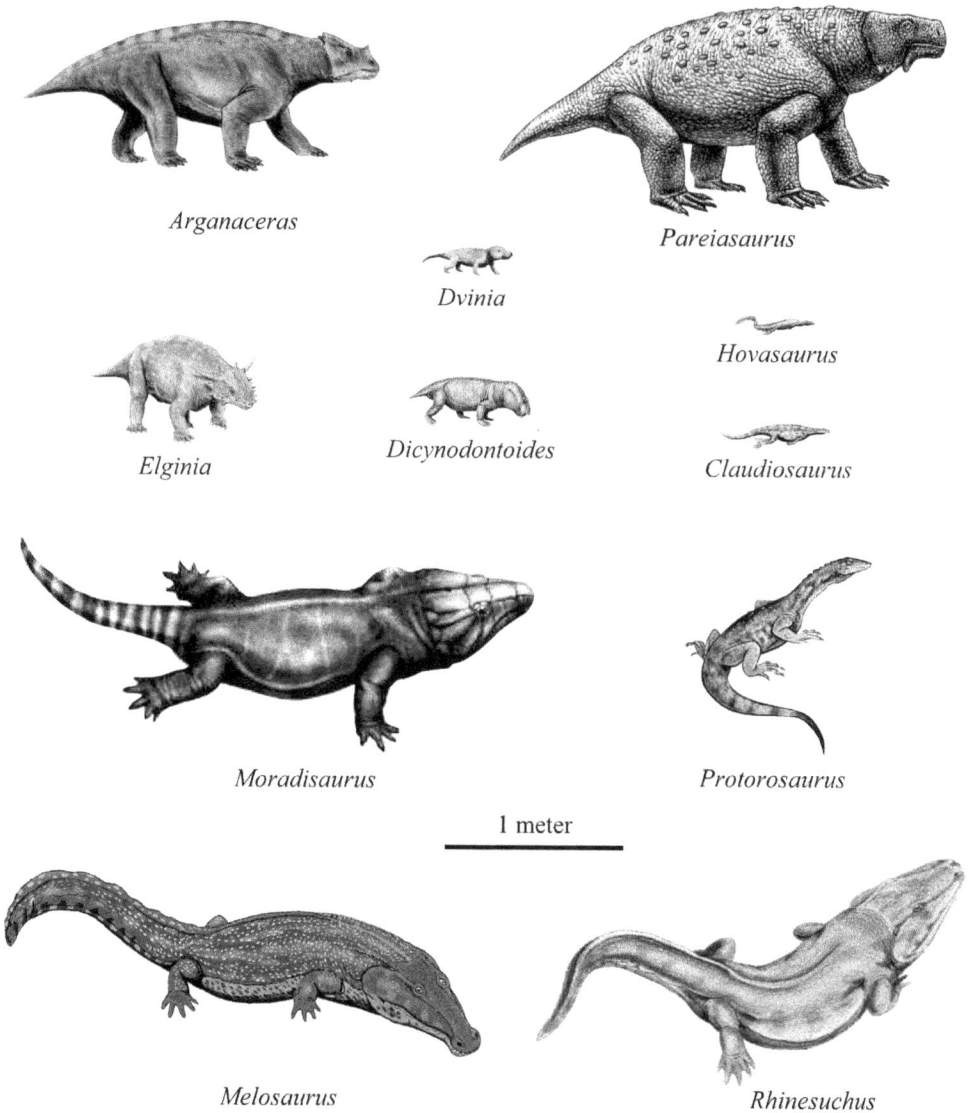

Arganaceras

Dvinia

Pareiasaurus

Hovasaurus

Elginia

Dicynodontoides

Claudiosaurus

Moradisaurus

Protorosaurus

1 meter

Melosaurus

Rhinesuchus

Figure 42 — Late Permian animals that were common outside the Virginia region. Arganaceras, Pareiasaurus, *and* Elginia *were parareptilian pareiasaurs;* Hovasaurus *and* Claudiosaurus *were neodiapsids;* Moradisaurus *was a captorhinid;* Protorosaurus *was an archosaur; and* Melosaurus *and* Rhinesuchus *were temnospondyl amphibians. Animals close to* Dvinia, *a cynodont, and* Dicynodontoides, *a dicynodont, survived the end of the Paleozoic and gave rise to the mammal-like reptiles that lived in Virginia during the Triassic Period of the Mesozoic. See Appendix 2 for the sources of this picture.*

CHAPTER

11

WHERE TO SEE PALEOZOIC FOSSILS AND EXHIBITS RELATED TO VIRGINIA

The Paleozoic history of vertebrate life encompasses more time than all of the remaining parts of Earth's history. Even so, there is relatively little on exhibit to see in the Virginia region that gives the public any meaningful sense of what transpired during this long and eventful interval of time. One significant exception is the "Fossil Overlook Exhibit" at the Virginia Museum of Natural History in Martinsville, Virginia, which includes a nice cast display of the skull and neck armor of the Devonian placoderm fish (*Dunkleosteus)* as well as a scaled flesh model of its entire anatomy. This huge fish is sure to impress (and perhaps terrify) all who contemplate its large size and razor jaws.

To get other meaningful glimpses of this early part of Earth's history in nearby areas, one must venture out of state. In Washington, D.C., the Smithsonian Institution has impressive new exhibits that opened in June of 2019. These include a few exhibits on the early history of vertebrate life on a global scale. Farther afield to the north, the Academy of Natural Sciences of Drexel University in Philadelphia, Pennsylvania, and the New Jersey State Museum in Trenton, New Jersey, both have nice exhibits that include specimens from the Paleozoic that are worth seeing. The Carnegie Museum in Pittsburgh, Pennsylvania, also has material from the Appalachian region that makes a visit worthwhile to its Benedum Hall of Geology.

In the state of New York, excellent, wide-ranging exhibits at the American Museum of Natural History in New York City are well worth visiting. There are also nice Paleozoic exhibits to be seen at the Museum of the Earth in Ithaca. Specifically of a Paleozoic flavor, the Gilboa Museum of New York (www.gilboafossils.org) specializes in the Devonian history of the New York state region. There, visitors can see Devonian animal fossils and also remnants of the Gilboa Forest, which is one of the oldest known fossil forest sites in the world. Near this site, at Gilboa Dam, there is an exhibit of large upright Devonian tree stumps to be seen.

While far afield for most Virginia visitors, very good fossil exhibits of Paleozoic creatures can be found in Nova Scotia, Canada. There, the Museum of Natural History in Halifax has large slabs of rock on exhibit with trackways of giant millipedes on them. Similarly, the Joggins Fossil Cliffs (a UNESCO World Heritage Site) has fascinating fossils from the Carboniferous time period that also are well worth a visit. Another excellent series of exhibits worth visiting can be seen at the Royal Ontario Museum in Toronto, Canada, at the Willner Madge Gallery "Dawn of Life" exhibit, which opened in 2021. These exhibits tend to have a rather global perspective, but even so, they provide an exceptional sense of what the pre-Cambrian and Paleozoic world was like and the remarkable array of strange creatures that once inhabited this world long ago. While a long way from the Virginia region, these parts of Canada also have many other fascinating things to see and do. A trip there as an extended family adventure would be well worth the time and cost of such a visit.

REFERENCES CITED

Argencia Sinc, 2013, A new species of marine fish that lived in the Devonian: https://todropscience.tumblr.com/post/51734353510/a-new-specie-of-marine-fish-that-lived-in-the-devonian. [*Machaeracanthus*]

Arkle, T., Jr., Beissell, D. R., Larese, R. E., Nuhfer, E. B., Patchen, D. G., Smosna, R. A., Gillespie, W. H., Lund, R., Norton, W. and Pfefferkorn, H. W., 1979, West Virginia and Maryland *in* The Mississippian and Pennsylvanian (Carboniferous) Systems in the United States: *United States Geological Survey Professional Paper*, no. 1110, p. D1-D35. [*Chomatodus, Cladodus, Deltodus, Desmiodus, Gyracanthus*, aff. *Helodus, Hybocladodus, Janassa, "Peltodus," Peripristis, Petalodus, Physonemus* (= *Falcatus*), *Poecilodus, Psephodus, Tranodis, Venustodus*]

Bailey, C. M. and Peters, S. E., 1998, Glacially influenced sedimentation in the late Neoproterozoic Mechum River Formation, Blue Ridge Province, Virginia: *Geology*, vol. 26, no. 7, p. 623-626.

Baird, D., 1957, A *Physonemus* spine from the Pennsylvanian of West Virginia: *Journal of Paleontology*, vol. 31, no. 5, p. 1010-1018.

Baird, D., 1978, Studies on Carboniferous freshwater fishes: *American Museum Novitates*, no. 2641, p. 1-22. [*Bandringa*, aff. *Commentrya, Elonichthys, Haplolepis, Rhabdoderma, Rhizodopsis, Xenacanthus*]

Beerbower, J. R., 1963, Morphology, paleoecology, and phylogeny of the Permo-Pennsylvanian amphibian *Diploceraspis*: *Bulletin of the Museum of Comparative Zoology*, vol. 130, p. 31-108.

Berman, D. S., Henrici, A. C., Brezinski, D. K. and Kollar, A. D., 2010, A new trematopid amphibian (temnospondyli: dissorophoidea) from the Upper Pennsylvanian of western Pennsylvania: earliest record of terrestrial vertebrates responding to a warmer, drier climate: *Annals* of the Carnegie Museum, vol. 78, no. 4, p. 289-318. [*Fedexia*]

Bernardi, M., Petti, F. M., Kustatscher, E., Franz, M., Hartkopf-Fröder, C., Labandeira, C. C., Wappler, T., van Konijnenburg-van Cittert, J. H., Peecook, B. R. and Angielczyk, K. D., 2017, Late Permian (Lopingian) terrestrial ecosystems: a global comparison with new data from the low-latitude Bletterbach Biota: *Earth-Science Reviews*, vol. 175, p. 18-43.

Branson, E. B., 1910, Amphibian footprints from the Mississippian of Virginia: *Journal of Geology*, vol. 18, p. 356-358. [*Dromopus aduncus*]

Butts, C., 1940, Geology of the Appalachian Valley of Virginia: *Virginia Geological Survey Bulletin*, vol. 52, p. 114 and 179. [*Machaeracanthus*]

Butts, C. and Edmundson, R. S., 1966, Geology and Mineral Resources of Frederick County: *Virginia Division of Mineral Resources*, Bulletin 80, 142 pp. [*Eczematolepis* (=*Acantholepis*)]

Caldwell, W. and Wellstead, C., 2007, Additions to the Middle Pennsylvanian Ash Branch fossil vertebrate fauna: *Proceedings of the West Virginia Academy of Science*, vol. 79, no. 1, p. 25. [*Ageleodus, Megalichthys, Onychodus, Orcanthus, Sagenodus*]

Carroll, R. L., Bossy, K. A., Milner, A. C., Andrews, S. M. and Wellstead, C. F., 1998, *Handbuch der Palaoherpetologie / Encyclopedia of Paleoherpetology*: Teil 1 / Part 1 Lepospondyli: Microsauria, Nectridea, Lysorophia, Adelospondyli, Aistopoda, Acherontiscidae, 216 pp. [*Brachydectes, Diplocaulus, Diploceraspis, Lysorophus, Molgophis, Odonterpeton, Oestocephalus, Phlegethontia, Pleuroptyx, Ptyonius, Sauropleura, Tuditanus*]

Cope, E. D., 1892, On some new and little known Paleozoic vertebrates: *Proceedings of the American Philosophical Society*, vol. 30, p. 221-229. [*Bothriolepis, Ganorhynchus, Holonema, Holoptychius, Osteolepis*]

Cope, E. D., 1897, On New Paleozoic Vertebrata from Illinois, Ohio and Pennsylvania: *Proceedings of the American Philosophical Society*, vol. 36, p. 71-91. [*Holptychus*]

Daeschler, E. B. and Cressler, W. L., III, 2011, Late Devonian paleontology and paleoenvironments at Red Hill and other fossil sites in the

Catskill Formation of north-central Pennsylvania, *in* Ruffolo, R. M., and Ciampaglio, C. N., (eds.), From the Shield to the Sea: Geologic trips from the 2011 joint meeting of the GSA Northeastern and North-Central sections: *Geological Society of America*, Field Guide 20, 16 pp. [*Ageleodus, Ctenacanthus, Densignathus, Groenlandaspis, Gyracanthus, Hyneria, Hynerpeton, Limnomis, Phyllolepis, Sauripterus, Turrisaspis,* cf. Whatcheeridae]

Denison, R. H., 1964, The Cyathaspididae; a family of Silurian and Devonian jawless vertebrates: Fieldiana, *Geology*, vol. 13, no. 5, p. 309-473. [*Americaspis, Vernonaspis*]

Eastman, C. R., 1908, Devonian fishes of Iowa: *Iowa Geological Survey*, vol. 18, p. 29-363. [*Apedodus, Bothriolepis, Cladodus, Coccosteus, Ctenacanthus, Dinichthys, Dipterus, Glyptopomus, Granorhynchus, Gyracathus, Heliodus, Helodus, Holonema, Holoptychius, Homacanthus, Phyllolepis, Sauripteris, Spenophorus*]

Elliott, D. K. and Petriello, M. A., 2011, New Poraspids (Agnatha, Heterostraci) from the Early Devonian of the western United States: *Journal of Vertebrate Paleontology*, vol. 31, no. 3, p. 518-530.

Evans, S. D., Tu, C.-Y., Rizzo, A. and Droser, M. L., 2022, Environmental drivers of the first major animal extinction across the Ediacaran White Sea-Nama transition: *Proceedings of the National Academy of Sciences*, vol. 119, no. 46, p. e2207475119.

Eyles, N. and Young, G., 1994, Geodynamic controls on glaciation in Earth history, p. 10-18 *in* Deynoux, M., Miller, J. M. G., Domack, E. W., Eyles, N., Fairchild, I. J., and Young, G. M. (eds.): *Earth's Glacial Record (no. 5)*, Cambridge University Press, 288 pp.

Fenton, C. L. and Fenton, M. A., 1958, *The Fossil Book: A record of prehistoric life*: Doubleday and Company Incorporated, Garden City, New York, 482 pp.

Garton, E. R., 1978, Fossil fishes from the Grayden Shale, Fayette County, West Virginia: *Proceedings of the West Virginia Academy of Sciences*, vol. 50, p. 38 (abstract). [*Sagenodus*]

Gathright, T. M., II, 1976, Geology of the Shenandoah National Park, Virginia: *Virginia Division of Mineral Resources Bulletin*, no. 86, 93 pp.

Giffin, E. B., 1979. Silurian vertebrates from Pennsylvania: *Journal of Paleontology*, vol. 53, no. 2, p. 438-445. [*Logania, Thelodus*]

Grazhdankin, D., 2004, Patterns of distribution in the Ediacaran biotas: facies versus biogeography and evolution: *Paleobiology*, vol. 30, no. 2, p. 203-221.

Greb, S. F., Pashin, J. C., Martino, R. L. and Eble, C. F., 2008, Appalachian sedimentary cycles during the Pennsylvanian: Changing influences of sea level, climate, and tectonics, *in* Fielding, C. R., Frank, T. D., and Isbell, J. L., (eds.), Resolving the Late Paleozoic ice age in time and space: *Geological Society of America*, Special Paper 441, p. 235-248.

Harris, A. G., Stamm, N. R., Weary, D. J., Repetski, J. E., Stamm, R. G. and Parker, R. A. 1994, Conodont color alteration index (CAI) map and conodont-based age determinations for the Winchester 30' x 60' Quadrangle and adjacent area, Virginia, West Virginia, and Maryland: *United States Geological Survey*, Miscellaneous Field Studies Map 2239, 40 pp. [*Acanthodus, Acontiodus, Adetognathus, Amorphognathus, Ancyrodella, Ansella, Appalachignathus, Bellodella, Baltoniodus, Cambrooistodus, Cavusgnathus, Chosonodina, Clavohamulus, Cordylodus, Curtognathus, Dapsilodus, Decoriconus, Diaphorodus, Distomodus, Drepanodus, Drepanoistodus, Dvorakia, Eoconodontus, Erismodus, Eucharodus, Glyptoconus, Gnathodus, Hindeodus, Hirsutodontus, Iapetognathus, Icriodus, Juanognathus, Kladognathus, Leptochirognathus, Ligonodina, Lochriea, Loxodus, Macerodus, Mehlina, Mesotaxis, Milaculum, Monocostodus, Neomultioistodus, Oepikodus, Oistodus, Oneotodus, Oulodus, Ozarkodina, Panderodus, Pandorinellina, Paraprionodus, Pedovis, Periodon, Phakelodus, Phragmodus, Plectodina, Polygnathus, Proconodontus, Prooneotodus, Protopanderodus, Pseudooneotodus, Pteracontiodus, Rossodus, Scandopus, Scolopodus, Semiacontiodus, Teridontus, Tropodus, Ulrichodina, Utahconus, Variabiloconus, Vogelgnathus, Westergaardodina*]

Haubold, H., 1971, Ichnia Amphibiorum et Reptiliorum Fossilium, *in* Kuhn, Oskar (ed.), *Encyclopedia of Paleoherpetology*: Gustav Fischer Verlag, Stuttgart, Part 18, 124 pp. [*Dromopus aduncus* is here moved to *Asperipes aduncus,* but later it was moved back to *Dromopus aduncus* by Lucas (2019)]

Heckel, P. H., Barrick, J. E. and Rosscoe, S. J., 2011, Conodont-based correlation of marine units in Lower Conemaugh Group (Late Pennsylvanian) in Northern Appalachian Basin: *Stratigraphy*, vol. 8, no. 4, p. 253-269.

Helfrich, C. T., 1975, Silurian conodonts from Wills Mountain anticline, Virginia, West Virginia, and Maryland: *Geological Society of America Special Paper*, vol. 161, 82 pp. [*Hindeodella, Ligonodina, Lonchodina, Neoprioniodus, Ozarkodina, Panderosus, Plectospathodus, Spathognathodus, Trichonodella*]

Helfrich, C. T., 1978, A conodont fauna from the Keyser Limestone of Virginia and West Virginia: *Journal of Paleontology*, vol. 52, no. 5, p. 1133-1142. [*Belodella, Delotaxis, Icriodus, Ozarkodina, Panderodus*]

Helfrich, C. T., 1980, Late Llandovery-early Wenlock conodonts from the upper part of the Rose Hill and the basal part of the Mifflintown formations, Virginia, West Virginia, and Maryland: *Journal of Paleontology*, vol. 54, no. 3, p. 557-569. [*Aspidognathus, Carniodus, Distomodus, Kockelella, Ozarkodina, Panderodus, Pseudooneotodus, Pterospathodus, Walliserodus*]

Hook, R. W. and Baird, D., 1988, An overview of the Upper Carboniferous fossil deposit at Linton, Ohio: *Ohio Journal of Science*, vol. 88, no. 1, p. 55-60. [*Amphibamus, Anthracodromeus, Archaeothyris, Cephalerpeton, Cocytinus, Colosteus, Conchopoma, Ctenerpeton, Diceratosaurus, Elonichthys, Erpetosaurus, Eusauropleura, Gandrya, Haplolepis, Leptophractus, Macrerpeton, Megalocephalus, Microhaplolepis, Molgophis, Odonterpeton, Ophiderpeton, Orthacanthus, Parahaplolepis, Phlegethontia, Pleuroptyx, Ptyonius, Pyritocephalus, Raphetes, Rhabdoderma, Sagenodus, Saurerpeton, Sauropleura, Stegops, Tuditanus, Xenacanthus*]

Hook, R. W. and Baird D., 1993, A new fish and tetrapod assemblage from the Allegheny Group (Late Westphalian, Upper Carboniferous) of eastern Ohio, U.S.A. *in* Heidtke, U. (compiler), New research on Permo-Carboniferous faunas: *Pollichia-Buch*, vol. 29, p. 143-154. [*Sagenodus*]

Hower, J. C., O'Keefe, J. M. and Eble, C. F., 2008, Tales from a distant swamp: Petrological and paleobotanical clues for the origin of the sand coal lithotype (Mississippian, Valley Fields, Virginia): *International Journal of Coal Geology*, vol. 75, no. 2, p. 119-126.

Huang H., Cawood P. A., Hou, M., Yang, J., Ni, S., Du, Y., Yan, Z. and

Wang, J., 2016, Silicic ash beds bracket Emeishan large igneous province to 1 m. y. at ~260 Ma: *Lithos*, vol. 264, p. 17-27.

Johnson, T. A., Vervoort, J. D., Ramsey, M. J., Southworth, S. and Mulcahy, S. R., 2020, Tectonic evolution of the Grenville Orogen in the central Appalachians: *Precambrian Research*, vol. 346, p. 1-21.

Kissel, R., 2010, Morphology, Phylogeny, and Evolution of Diadectidae (Cotylosauria: Diadectomorpha): University of Toronto, Ph. D. dissertation, 185 pp. [*Ambedus, Ctenospondylus, Sagenodus, Trimerorhachis*]

Leutze, W. P., 1960, Silurian fish fossils in the Salina Basin: *Bulletin of the Geological Society of America*, vol. 71, p. 215-218. [Listed in this paper as Heterostraci indeterminate, but at least in part now included within *Vernonaspis*]

Lucas, S. G., 2013, Vertebrate biostratigraphy and biochronology of the Upper Paleozoic Dunkard Group, Pennsylvania-West Virginia-Ohio, USA: *International Journal of Coal Geology*, v. 119, p. 79-87.

Lucas, S. G., 2015, *Thinopus* and a critical review of Devonian tetrapod footprints: *Ichnos*, vol. 22, p. 136-154.

Lucas, S. G., 2019, An ichnological perspective on some major events of Paleozoic tetrapod evolution: *Bolletino della Societá Paleontologica Italiana*, vol. 58, no. 3, p. 223-266. [*Baropus, Batrachichnus, Dimetropus, Dromopus, Hylopus, Limnopus, Matthewichnus, Palaeosauropus, Pseudobradypus*]

Lucas, S. G., Kollar, A. D., Berman, D. S. and Henrici, A. C., 2016, Pelycosaurian-grade (amniota: synapsida) footprints from the Lower Permian Dunkard Group of Pennsylvania and West Virginia: *Annals of the Carnegie Museum*, vol. 83, no. 4, p. 287-294. [*Acheloma, Ctenospondylus, Diadectes, Dimetrodon, Dimetropus, Diploceraspis, Dromopus, Edaphosaurus, Eryops, Lysorophus, Megamolgophis, Ophiacodon, Protorothyrus*]

Lund, R., 1972a, Notes on the vertebrate fossils of the Elm Grove area, West Virginia: *West Virginia Geological and Economic Survey*, I. C. White Memorial Symposium Field Trip (September 27, 28, 29 1972), p. 51. [*Diploceraspis*, aff. *Edops, Lysorophus*]

Lund, R., 1972b, Upper Washington Limestone and Lower Greene Formation near the axis of the Nineveh syncline, Vertebrate fossils from the Washington Stone Company Quarry, Vance, Pa.: *West Virginia Geological and Economic Survey*, I. C. White Memorial Symposium Field Trip (September 27, 28, 29 1972), p. 56-58. [*Dimetrodon, Diploceraspis, Ectosteorhachis, Edaphosaurus, Hybodus, Lysorophus, Monongahela, Orthacanthus, Sagenodus, Xenacanthus*]

Lund, R., 1985, The morphology of *Falcatus falcatus* (St. John and Worthen), a Mississippian stethacanthid chondrichthyan from the Bear Gulch Limestone of Montana: *Journal of Vertebrate Paleontology*, vol. 5, no. 1, p. 1-19. [*Falcatus* replaces *Physonemus*]

Marchetti, L., Klein, H., Buchwitz, M., Ronchi, A., Smith, R. M., De Klerk, W. J., Sciscio, L. and Groenewald, G. H., 2019, Permian-Triassic vertebrate footprints from South Africa: Ichnotaxonomy, producers and biostratigraphy through two major faunal crises: *Gondwana Research*, vol. 72, p.139-168.

Marsh, O. C., 1896, Amphibian footprints from the Devonian: *American Journal of Science*, vol. 152, p. 374-375. [*Thinopus*]

Martin, W. D., 1998, Geology of the Dunkard Group (Upper Pennsylvanian-Lower Permian) in Ohio, West Virginia, and Pennsylvania: *Ohio Division of Geological Survey*, bull. 73, p. 1-49. [*Dimetrodon, Edaphosaurus, Eryops, Ophiacodon*]

McMenamin, M. A. S. and Weaver, P. G., 2004, Middle Cambrian polymeroid trilobites and correlation of the Carolina and Augusta terranes: *Southeastern Geology*, vol. 43, no. 1, p.21-38.

Mickle, K. E., 2018, A new lower actinopterygian fish from the Upper Mississippian Bluefield Formation of West Virginia, USA: *PeerJ*, 6: e5533, doi:10.7717/peerj.5533, pmid: 30186696. [*Bluefieldius*]

Miller, J. M. G., 1986, Upper Proterozoic glaciogenic rift-valley sedimentation: Upper Mount Rogers Formation, southwestern Virginia: *American Association of Petroleum Geologists*, bull. 70, no. 5, (CONF-860624-Journal ID: AAPGBS; ISSN 0002-7464).

Moran, W. E., 1952, Location and stratigraphy of known occurrences of fossil tetrapods in the Upper Pennsylvanian and Permian of Pennsylvania, West Virginia, and Ohio: *Annals of the Carnegie Museum*, vol.

33, p. 1-45. [*Brachydectes*]

Morris, S. C. and Caron, J. B., 2014, A primitive fish from the Cambrian of North America: *Nature*, vol. 512, no. 7515, p. 419-422. [*Metaspriggina*]

Mossman, D. and Grantham, R., 1999, Vertebrate trackways in the Parrsboro Formation (Upper Carboniferous) at Rams Head, Cumberland County, Nova Scotia: *Atlantic Geology*, vol. 35, no. 3, p. 186-196. [*Cursipes, Dromillopus, Hylopus, Pseudobradypus*]

Murphy, J. L., 1971, Eryopsid remains from the Conemaugh Group, Braxton County, West Virginia: *Southeastern Geology*, vol. 13, no.4, p. 265-273. [*Glaukerpeton*]

Narbonne, G. M., Saylor, B. Z. and Grotzinger, J. P., 2015, The youngest Ediacaran fossils from Southern Africa: *Journal of Paleontology*, vol. 71, no. 6, p. 953-967. [*Swartpuntia*]

Orndorff, R. C., 1988, Latest Cambrian and earliest Ordovician conodonts from the Conococheague and Stonehenge limestones of northwestern Virginia: *United States Geological Survey*, bull. 1837, p. A1-A18, 2 pls. [*Acodus, Acontiodus, Cambrooistodus, Clavohamulus, Cordylodus, Drepanoistodus, Eoconodontus, Fryxellodontus, Hirsutodontus, Loxodus, Oistodus, Paltodus, Proconodontus, Prooneotodus, Rossodus, Semiacontiodus, Teridontus, Utahconus*]

Petrychenko, Y., Peryt, T. M. and Chechel, E. I., 2005, Early Cambrian seawater chemistry from fluid inclusions in halite from Siberian evaporites: *Chemical Geology*, vol. 219, numbers 1-4, p. 149-161.

Rankin, D. W., 1970, Stratigraphy and structure of Precambrian rocks in northwestern North Carolina, p. 227-246 *in* Fisher, G. W., et al., (eds.), *Studies in Appalachian Geology: Central and Southern*: New York, John Wiley and Sons, Inc., 460 pp.

Rankin, D. W., 1992, The volcanogenic Mount Rogers Formation and the overlying glaciogenic Konnarock Formation; two Late Proterozoic units in southwestern Virginia: *United States Geological Survey Bulletin* 2029, 26 pp.

Rankin, D. W., Stern, T. W., Reed, J. C. and Newell, M. F., 1969, Zircon ages of felsic volcanic rocks in the Upper Precambrian of the Blue

Ridge, Appalachian Mountains: *Science*, vol. 166, no. 3896, p. 741-744.

Raymond, P. E., 1911, A preliminary list of the fauna of the Allegheny and Conemaugh series in western Pennsylvania: *Report of the Pennsylvania Topographical and Geological Survey Committee*, vol. 7, p. 83-98. [*Agassizodus, Cladodus, Deltodus, Dittodus, Fissodus, Petalodus*]

Reichow, M. K., Pringle, M. S., Al'Mukhamedov, A. I., Allen, M. B., Andreichev, V. L., Buslov, M. M., et al., 2009, The timing and extent of the eruption of the Siberian Traps large igneous province: implications for the end-Permian environmental crisis: *Earth and Planetary Science Letters*, vol. 277, nos. 1-2, p. 9-20.

Repetski, J. E. and Henry, T. W., 1983, A Late Mississippian conodont faunule from area of proposed Pennsylvanian System stratotype, eastern Appalachians: *Fossils and Strata*, vol. 15, p. 169-170. [*Adetognathus, Cavusgnathus, Gnathodus*]

Repetski, J. E. and Perry, W. J., Jr., 1980, Early Ordovician conodonts from the Bane dome, Giles County, Virginia: *United States Geological Survey Open-File Report*, no. 80-372, p. 1-11. [*Acanthodus,* cf. *Acodus, Acontiodus, Chosonodina, Clavohamulus, Drepanodus, Juanognatuthus?,* cf. *Loxodus, Paltodus, Paroistodus?, Scolopodus, Scolopodus, Scolopodus, Urichodina*]

Robbins, E. I. and Weems, R. E., 1988, Preliminary analysis of unusual palynomorphs from the Taylorsville and Deep Run Basins in the eastern Piedmont of Virginia: *United States Geological Survey Bulletin*, no. 1776, p. 40-57.

Rogers, W. B., 1882, The fossils of formation No. III in Virginia: *The Virginias*, vol. 3, p. 175.

Romer, A. S., 1933, Eurypterid influence on vertebrate history: *Science*, vol. 78, p. 114-117.

Romer, A. S., 1952, Late Pennsylvanian and Early Permian Vertebrates of the Pittsburgh-West Virginia Region: *Annals of the Carnegie Museum*, vol. 33, p. 47-113. [*Acanthodes, Agassizodus, Amblypterus, Baropus, Cladodus, Deltodus, Desmatodon, Dimetropus, Diploceraspis, Dittodus, Dryops, Edaphosaurus, Fissodus, Glaukerpeton, Limnosceloides, Lysorophus, Megamolgophis, Melanothyris* (= *Protorothyris*)*, Peripristis,*

Petalodus, Protorothyris, Sagenodus, Saurerpeton]

Sansom, I. J. and Andreev, P., 2017, The Ordovician enigma: fish, first appearances and phylogenetic controversies, *in* Johanson, Z., Richter, M., and Underwood, C., (eds.), *Evolution and Development of Fishes*: Cambridge University Press, Chapter 3, p. 59-69.

Sansom, I. J., Smith, M. P., Smith, M. M. and Turner, P., 1997, *Astraspis*, the anatomy and histology of an Ordovician fish: *Palaeontology*, vol. 40, no. 3, p. 625-643.

Saunders, A., and Reichow, M., 2009, The Siberian Traps and the End-Permian mass extinction: a critical review: *Chinese Science Bulletin*, vol. 54, no. 1, p. 20-37.

Schneider, J. W., Lucas, S. G. and Barrick, J. E., 2013, The Early Permian age of the Dunkard Group, Appalachian basin, U.S.A., based on spiloblattinid insect biostratigraphy: *International Journal of Coal Geology*, v. 119, p. 88-92.

Schultze, H.-P. and Chorn, J., 1997, The Permo-Carboniferous genus *Sagenodus* and the beginning of modern lungfish: *Contributions to Zoology*, vol. 67, no. 1, p. 9-70.

Schwab, F. L., 1976, Depositional environments, provenance, and tectonic framework: Upper part of the Late Precambrian Mount Rogers Formation, Blue Ridge Province, Southwestern Virginia: *Journal of Sedimentary Petrology*, vol. 46, no. 1, p. 2-13.

Selden, P. A., 1984, Autecology of Silurian eurypterids *in* M. G. Bassett and J. D. Lawson (eds.), Autecology of Silurian Organisms: *Special Papers in Palaeontology*, vol. 32, p. 39-54.

Smith, M. P., Sansom, I. J. and Repetski, J. E., 1996, Histology of the first fish: *Nature*, v. 380, no. 6576, p. 702-704. [*Anatolepis*]

Snyder, D., Turner, S., Burrow, C. J. and Daeschler, E. B., 2017, "*Gyracanthus*" *sherwoodi* (Gnathostomata, Gyracanthidae) from the Late Devonian of North America: *Proceedings of the Academy of Natural Sciences of Philadelphia*, vol. 165, p. 195-219.

Stamm, R. G. and Wardlaw, B. R., 2003, Conodont faunas of the late Middle Pennsylvanian (Desmoinesian) lower Kittanning cyclothem, U.S.A.: *Society of Economic Paleontologists and Mineralogists Special*

Publication, no. 77, p. 95-121. [*Hindeodus, Idiognathodus, Idioprioniodus*]

Südkamp, W. H. and Burrow, C. J., 2007, The acanthodian *Machaeracanthus* from the Lower Devonian Hünsruck Slate of the Hünsruck region (Germany): *Paläontologische Zeitschrift*, vol. 81, no. 1, p. 97-104.

Sundberg, F. A., Bennington, J. B., Wizevich, M. C. and Bambach, R. K., 1990, Upper Carboniferous (Namurian) amphibian trackways from the Bluefield Formation, West Virginia, USA: *Ichnos*, v. 1, p. 111-124. [*Hylopus, Proterogyrinus*]

Swartz, C. K., 1913, Vertebrata, *in* Middle and Upper Devonian: *Maryland Geological Survey*, The Lord Baltimore Press, p. 700-701. [*Glyptaspis* (= *Holonema*)]

Swartz, C. K., Prouty, W. F., Ulrich, E. O. and Bassler, R. S., 1923, *Silurian*: Johns Hopkins Press, vol. 8, 749 pp, pls. 1-67. [*Palaeaspis americana* in this volume is now called *Americaspis americana;* Denison (1964) implies but does not state that this material is probably *Vernonaspis*]

Terada, K., Morota, T. and Kato, M., 2020, Asteroid shower on the Earth-Moon system immediately before the Cryogenian period revealed by KAGUYA: *Nature Communications*, vol. 11, no. 1, p. 1-10.

Tian, Q., Zhao, F., Zeng, H., Zhu, M. and Jiang, B., 2022, Ultrastructure reveals ancestral vertebrate pharyngeal skeleton in yunnanozoans: *Science*, vol. 377, no. 6602, p. 218-222.

Tollo, R. P., Bailey, C. M., Borduas, E. A. and Aleinikoff, J. N., 2004, Mesoproterozoic geology of the Blue Ridge, *in* Southworth, S. and Burton, W., (eds.), *Geology of the National Capital Region*, Field Trip Guidebook 1264, chapter 2, p. 17-76.

Tomescu, A. M., Rothwell, G. W. and Honegger, R., 2006, Cyanobacterial macrophytes in an Early Silurian (Llandovery) continental biota: Passage Creek, Lower Massanutten Sandstone, Virginia, USA: *Lethaia*, vol. 39, no. 4, p. 329-338.

Tomescu, A. M., Rothwell, G. W. and Honegger, R., 2009, A new genus and species of filamentous microfossil of cyanobacterial affinity from Early Silurian fluvial environments (Lower Massanutten Sandstone,

Virginia, USA): *Botanical Journal of the Linnean Society*, vol. 160, no. 3, p. 284-289.

Tverdokhlebov, V. P., Tverdokhlebova, G. I., Minikh, A. V., Surkov, M. V. and Benton, M. J., 2005, Upper Permian vertebrates and their sedimentological context in the South Urals, Russia: *Earth-Science Reviews*, vol. 69, no. 1-2, p. 27-77.

Walcott, C. D., 1921, Cambrian brachiopoda: *United States Geological Survey Monograph*, vol. 51, 812 pp.

Weaver, P. G., McMenamin, M. A. S. and Tacker, R. C., 2006a, Paleoenvironmental and paleobiogeographical implications of a new Ediacaran body fossil from the Neoproterozoic Carolina Terrane, Stanly Co., North Carolina: *Precambrian Research*, vol. 105, p. 123-135.

Weaver, P. G., Tacker, R. C., McMenamin, M. A. S. and Webb, R. A., 2006b, Ediacaran body fossils of south-central North Carolina: Preliminary Report *in* Bradley, P. J. and Clark, T. W. (eds.) The Geology of the Chapel Hill, Hillsborough and Efland 7.5-Minute Quadrangles, Orange and Durham Counties, Carolina Terrane, North Carolina: *Carolina Geological Society Field Trip Guidebook*, p. 35-42.

Weaver, P. G., Tacker, R. C., McMenamin, M. A. S., Ciampaglio, C. N. and Webb, R. A., 2008, Additional Ediacaran body fossils of south-central North Carolina: *Southeastern Geology*, vol. 45, no.4, p. 225-232.

Weems, R. E., 1993, Stratigraphic Distribution and Bibliography of Fossil Fish, Amphibians, and Reptiles from Virginia: *United States Geological Survey Open-File Report*, no. 93-222, 49 pp. [*Asperipes, Bothriolepis*]

Weems, R. E., 2004, *Bothriolepis virginiensis*, a valid species of placoderm fish separable from *Bothriolepis nitida*: *Journal of Vertebrate Paleontology*, vol. 24, no. 1, p. 245-250.

Weems, R. E., 2022, Fossil footprints from the Early Permian Dunkard Group of Maryland, *in* Lucas, S. G. et al., (eds.), Fossil Record 8: *New Mexico Museum of Natural History and Science Bulletin*, no. 90, p. 461-465. [*Batrachichnus, Dromopus*]

Weems, R. E. and Grimsley, G. J., 2022, An Early Silurian fish from the Clinch (Tuscarora) Sandstone of Virginia: *The Mosasaur*, vol. 12, p.

97-103. [aff. *Tremataspis*]

Weems, R. E. and Lucas, S. G., 2021, The first record of vertebrate ichnofossils from the Norton Formation (Pennsylvanian, Middle Moscovian) in southwestern Virginia, U.S.A., *in* Lucas, S. G., Hunt, A. P., and Lichtig, A. J., (eds.), Fossil Record 7: *New Mexico Museum of Natural History and Science Bulletin*, no. 82, p. 497-504. [*Batrachichnus, Characichnos, Limnopus*, Tetrapoda indet.]

Weems, R. E. and Robbins, E. I., 2023, The stratigraphy and stratigraphic nomenclature of the Goochland Terrane in the Piedmont Province of east-central Virginia: *Stratigraphy*, vol. 20, no. 1, p. 39-58.

Weems, R. E. and Windolph, J. F., Jr., 1986, A new Actinopterygian fish (Paleonisciformes) from the Upper Mississippian Bluestone Formation of West Virginia: *Proceedings of the Biological Society of Washington*, vol. 99, no. 4, p. 584-601. [*Tanypterichthys*]

Weems, R. E., Beem, K. A. and Miller, T. A., 1981, A new species of *Bothriolepis* (Placodermi, Bothriolepidae) from the Upper Devonian of Virginia (U.S.A.): *Proceedings of the Biological Society of Washington*, vol. 94, no. 4, p. 984-1004.

Wellstead, C. F., 1991, Taxonomic revision of the Lysorophia, Permo-Carboniferous lepospondyl amphibians: *Bulletin of the American Museum of Natural History*, vol. 209, 90 pp. [*Brachydectes, Lysorophus*]

Werneburg, R. and Berman, D. S., 2012, Revision of the aquatic eryopid temnospondyl *Glaukerpeton avinoffi* Romer, 1952, from the Upper Pennsylvanian of North America: *Annals of the Carnegie Museum*, vol. 81, no. 1, p. 33-60.

Woodward, H. P., 1943, Devonian system of West Virginia: *West Virginia Geological Survey*, v. 15, 655 pp. [*Bothriolepis, Holoptychius*]

Zhu, Y. A., Li, Q., Lu, J., Chen, Y., Wang, J., Gai, Z., Zhao, W., Wei, G., Yu, Y., Ahlberg, P. E. and Zhu, M., 2022, The oldest complete jawed vertebrates from the early Silurian of China: *Nature*, vol. 609, no. 7929, p. 954-958. [*Shenacanthus, Xiushanosteus*]

APPENDIX 1
SUPERSCRIPT NOTES FROM TEXT

Chapter 1

1. The rocks at the surface of the Earth have been repeatedly recycled as the Earth has evolved. As a result, the original rocks that formed the Earth have been melted, recycled, and eroded to create the younger rocks that are in existence today. Because of this recycling process, the oldest rocks that still remain at the surface of the Earth are over four billion years old and are found in the core of the oldest known pieces of continental crust (for example, in the Acasta Gneisses and the Nuvvuagittuq Greenstone Belt in Canada). In Australia, recycled zircons in the Narryer Gneiss in the Jack Hills of Western Australia have been found to be 4.4 billion years old. The Earth and other rocky planets apparently were formed from large volumes of asteroids and meteors that were expelled from the Sun during its creation, and radiometric dating of some of these meteors shows that they mostly formed about 4.6 billion years ago. These provide us with our best estimate so far of the original time at which our Sun spun out the materials that came to make up our comets, asteroids, and planets including the Earth. A summary of the latest information on these oldest Earth rocks and meteors can be found at Wikipedia: en.wikipedia.org/wiki/Oldest_dated_rocks.

2. The likely origin of the world's oceans was from the heating and degassing of the primitive Earth as it was evolving into its present form. In the early part of the Earth's formation, the heat from the radioactive decay of elements that were abundant in the early Earth plus the heat liberated from the sinking of Earth's heavy minerals to form the core of the Earth together kept the world extremely hot, so that it was unlikely that our planet was cool enough to hold liquid water at its surface until 3.8 billion years ago or even later. For a more detailed discussion of this

matter, see "Why do we have an ocean?": https://oceanservice.noaa.gov/facts/why_oceans.html.

3. Fossil evidence for life on Earth goes back to at least 3.5 billion years ago, and this life could have originated in the early oceans of the Earth through a series of complex biochemical processes. Alternatively, though, life could have come to Earth from elsewhere in a large asteroid or meteor and taken up residence here in our early lifeless oceans. Either hypothesis is plausible based on what little we know so far. It is worth noting, however, that the DNA of all life on Earth today is fundamentally the same, strongly suggesting that what remains came from a common ancestor. Whether this ancestor was the only early life on Earth, or whether it was simply the sole survivor of an early competition between different life forms of different origins, remains unknown. See the Wikipedia article on "Abiogenesis" for a more detailed discussion of this subject: https://en.wikipedia.org/wiki/Abiogenesis.

4. The chemistry of the Earth's atmosphere has changed significantly throughout its long history. Three major compositional phases of Earth's atmosphere are presently recognized. The initial atmosphere consisted of gasses derived from the solar nebula, similar to the atmospheres that today envelop the planets Jupiter, Saturn, Uranus, and Neptune. It consisted primarily of hydrogen with lesser amounts of water vapor, methane, and ammonia. This atmosphere was eventually replaced by one that formed due to the outgassing of the Earth's interior as it heated up and began to carry interior material upward due to the initiation of plate tectonics. Late heavy bombardment of the Earth by huge asteroids also added similar gasses to this second Earth's atmosphere, which came to consist mostly of nitrogen, carbon dioxide, and chemically inert gasses such as neon and argon. By 3.4 billion years ago, nitrogen had come to form the majority of the Earth's atmosphere. Oxygen was then abundant, but it was strongly combined with carbon and other elements and so was not found as a free component of the second Earth atmosphere. About 2.4 billion years ago, the appearance and spread of cellular plant life initiated the Great Oxygenation Event, which took most of the carbon dioxide out of the atmosphere due to the spread of photosynthesis throughout the seas of the world. This process resulted in the deposition of most of the carbon in the Earth's atmosphere within the sea floor, leaving the oxygen to accumulate until

it became about a fourth of the Earth's atmosphere. The atmospheric oxygen content of the Earth's atmosphere has risen and fallen to some degree since this event, but generally, it has remained near this proportion in our atmosphere. The abundant presence of oxygen in our atmosphere, and dissolved in our oceans, is what has led to the development of the abundant animal life on our planet. The details of this process can be found in Wikipedia at: https://en.wikipedia.org/wiki/Atmosphere_of_Earth.

5. Plate tectonics has been, and continues to be, one of the most important processes that has caused the history of the Earth to develop as it has. Strong heating of the Earth during its early history, due to the breakdown of its contained radioactive elements, allowed the heavier elements within the Earth to sink toward its core and the lighter elements to move upward toward the surface. This process in turn also released a lot of heat that further melted the rocks within the Earth and allowed them to settle out into three major discrete layers called the crust, the mantle, and the core. About 3 to 3.5 billion years ago, the rocks of the mantle had settled into a series of circular flow patterns that allowed heat to escape upward toward the surface of the Earth in a methodical manner. The motion of these flows caused the thin crustal rocks near the surface of the Earth to move about, following the flow of the underlying mantle plumes as they moved along beneath the crust. These motions are what have caused the continents in the past to move about, merging and splitting as they responded to changes in the flow pattern of the underlying mantle plumes. This process appears to be ongoing and will continue into our future for billions more years. Details of this process can be found in Wikipedia at: https://en.wikipedia.org/wiki/Plate_tectonics.

6. Several recent articles posted online state that the oldest known rocks in Virginia are 1.8 billion years old. This seems to be taken from a suggestion by Sinha and Bartholomew (1984) that some of the layered gneisses near Nellysford perhaps could be as old as 1.8 billion years. Later work, however, has not supported this suggestion. In all likelihood, the oldest discernable ages for the Grenville gneisses in the Blue Ridge region are in the range of 1.2 to 1.0 billion years (Christopher Bailey, personal communication, 2022). The most accessible of these rocks lie in the Blue Ridge region (Tollo et al., 2004; Johnson et al.,

2020), though the less easily visited State Farm Gneiss in the Accreted Terranes region (Figure 1) is of similar age (Weems and Robbins, 2023). The best places to easily see these oldest rocks are in some of the mountainsides exposed along the Skyline Drive and the Blue Ridge Parkway, for example at The Pinnacle, Oventop Mountain, and Old Rag Mountain (Gathright, 1976). These rocks formed during a major continental collision that thrust the Virginia region upward into a mountain belt probably equal in height to the Himalayan Mountains of today (Johnson et al., 2020).

7. In North America, the oldest known sedimentary rocks are in Greenland and are about 3.9 billion years old: https://www.lpi.usra.edu/education/timeline/gallery/slide_21.html. Somewhat younger sedimentary rocks, about 3.75 billion years old, also are known from Quebec, Canada: https://scitechdaily.com/diverse-life-forms-evolved-3-75-billion-years-ago-challenging-the-conventional-view-of-when-life-began. Slightly younger banded iron deposits (ca. 2.8 to 2.5 billion years old) are early sedimentary deposits preserved in the Great Lakes region in the northern United States and Canada. These deposits have been very important sources of iron ore in the past: https://en.wikipedia.org/wiki/Banded_iron_formation. These rocks provide a regional source for early sedimentary history in North America.

8. Remanent magnetism is the magnetic orientation of minerals that form or become magnetically oriented at the time that they are deposited in the geologic record. Most remanent magnetism is formed either in molten rocks as they cool in a sedimentary environment, or in sediment as small magnetic crystals settle out of their water column slowly enough for them to become oriented with the magnetic field of the Earth at the time they are settling. This magnetic orientation is aligned with the latitude of the environment from which the crystals are forming or settling but does not provide any information about the longitude where the crystals are settling out of their surroundings. Learning the age and the latitude of rocks within continental masses has allowed geologists to determine the latitude of the various continental masses at the time that these rocks were forming. For rocks younger than about 200 million years, the pattern of seafloor development and spreading provides a clear idea of where the various continental masses were located on the Earth at the time these rocks were forming. Before 200

Ma, however, the location of ancient longitudes is much more subjective and intuitive. Because the basaltic sea floors are being constantly recycled by plate tectonics, none of this seafloor rock is apparently any older than about 200 Ma. Only continental granitic rocks preserve ages that are far more ancient than this. For more information, see: https://en.wikipedia.org/wiki/Natural_remanent_magnetization.

9. This ancient island arc belt is today known as Avalonia, which is named in part for the Avalon Peninsula in Newfoundland, Canada, and as Carolinia, which is named for the Carolina region of the United States. Avalonia and Carolinia initially developed about 800 Ma as a volcanic island arc on the northern margin of Gondwana. Later, these areas rifted away from Gondwana and eventually collided with parts of what would eventually become America, Canada, and Western Europe. Because this island arc was an isolated land mass for many millions of years during the Cambrian Period, it developed a distinctive fauna that serves to identify the various fragments of this old island arc as a formerly isolated land and coastal shelf region. More on these ancient land masses can be found at: https://en.wikipedia.org/wiki/Avalonia and at: https://en.wikipedia.org/wiki/Carolina_terrane.

Chapter 2

10. This earliest known part of Virginia's prehistory has been documented by a number of articles over the past 80 years (for example: Rankin et al., 1969; Rankin, 1970, 1992; Schwab, 1976; Miller, 1986; and Bailey and Peters, 1998). The most restrictive age for this interval of time has been established from the Mechum River Formation, which formed between 730 and 700 million years ago (Bailey and Peters, 1998).

11. There has been much speculation as to why the Earth became so cold back at that time, but hard evidence remains elusive. The most interesting possibility that has been suggested is that there was an interstellar dust and meteor cloud that swept through our solar system around 800 Ma. This dust and debris cloud could have dimmed much of the heat from the Sun for a long period of time (Terada et al., 2020).

Chapter 3

12. For more information on this event, see Evans et al. (2022).

13. This locale and earlier literature have been recently discussed in Weaver et al. (2008).

14. There are differences in the various known Ediacaran faunas, but they seem to reflect variations in environmental settings and not differences in time. Over time, the Ediacaran faunas seem to have remained fairly consistent in terms of the fossils known from their assemblages (Grazh-dankin, 2004).

Chapter 4

15. The concentration of calcium in seawater has been shown to have un-dergone a great increase at the beginning of the Cambrian (e.g., Petry-chenko et al., 2005). This great increase in the concentration of calcium ions in seawater appears to strongly correlate with the appearance of calcium and phosphorus-rich shells in many of the Early Cambrian an-imals.

16. The latest article on this ancient vertebrate ancestor is by Tian et al. (2022).

17. This fossil has been most recently discussed by Morris and Caron (2014).

18. A useful and informative article on conodonts can be found at Wiki-pedia: https://en.wikipedia.org/wiki/Conodont. Typically, six types of elements are found in the body of conodont animals when they are preserved as impressions. This means that six "species" of conodonts are normally found within the body of a single biological animal species. There is no guarantee that all six types are going to be found in any sin-gle sample, but this is generally close to being true and thus provides a reasonable estimate of the actual number of biological species of animals that were present at any single locality that has been sampled.

Chapter 5

19. These localities are summarized in Sansom et al. (1997). Based on the places where they have been found, it seems that these fish inhabited only the shallow seas that lay upon the midcontinent region of Laurasia. No evidence of fish remains has been found so far in the Ordovician strata of Appalachia, which then were in distinctly deeper water than the midcontinent regions. The Cambrian, Lower Ordovician, and Middle Ordovician lithologies typically found in the ancient Appalachian sea-way are shallow-water carbonate deposits (e.g., Butts and Edmundson,

1966). Starting early in the Late Ordovician, however, the Appalachian region began to sink and its carbonate deposits became replaced with deep water shales called the Martinsburg Formation. This change happened because, in the Late Ordovician, the eastern North American sea was deepening as the Iapetus Sea began to approach a subduction zone. This change in the marine environment in western Virginia and related areas of the Appalachian seaway apparently made the Late Ordovician deep-water environments in the Virginia area inhospitable to the marine vertebrates which were beginning to spread across the shallow-water central parts of North America. Long ago, W. B. Rogers (1882) reported fish scales from the Martinsburg Formation in southern Virginia, but a search for his locality by Weems (1981) failed to turn up any sign of them in their designated area. Weems concluded that what Rogers had seen and reported were rhomboidal segments of graptolite fossils, which do occur at this locality in the Martinsburg. To date, no fish remains have been convincingly reported from any of the Ordovician strata in Virginia.

20. This idea was proposed by Romer (1933). Since then, Selden (1984) reported on the presence of unstructured masses containing disarticulated agnathan fish fragments in the Moncks Water fish bed in the Hagshaw Hills Silurian inlier in Scotland, which were found together with well-preserved fish and the eurypterid *Lanarkopterus doloichoscheles*. He interpreted these masses as eurypterid coprolites. More recently, Elliott and Petriello (2011) have described a fractured dorsal shield of *Lechriaspis patula* from the Early Devonian which has puncture marks from what they convincingly interpreted to be a eurypterid claw. These observations provide good evidence that eurypterids were important predators of fossil agnathan fish during the early Paleozoic.

21. The history of Ordovician fishes was recently discussed in Sansom and Andreev (2017). If readers would like more information on this early part of fish history beyond the Virginia region, this is an excellent place to start.

Chapter 6

22. The senior author has been fortunate to visit Taiwan and personally see this process of uplift and westward shedding of sediment from Taiwan across the South China Sea toward the mainland of China.

23. This fossil fish was described recently by Weems and Grimsley (2022).

24. A diversity of these early jawed fishes has recently been reported from China (Zhu et al., 2022). These authors note that molecular studies of modern jawed vertebrates suggest that the fossil record of these fishes may actually lie somewhat farther back in time in the Late Ordovician, but so far, no fossils of Ordovician jawed fish have been found.

25. Interestingly, some of the evidence for this earliest stage of the advance of plant life from the sea onto land has been documented from the Massanutten Syncline in northern Virginia (Tomescu et al., 2006, 2009).

Chapter 7

26. All but one of the kinds of Devonian fish shown in Table 1 (30 species) are from the Upper Devonian. A single species (*Machaeracanthus peracutus*) is known from the Lower Devonian.

27. This fish specimen was described by Butts (1940). It is primarily known from fossils of its prominent pectoral and pelvic fin spines (Figure 20), which apparently made it exceptionally stable for traveling in the waters of the open seas (Südkamp and Burrow, 2007).

28. A good summary of the Late Devonian fishes that were found long ago in the Pennsylvania region can be found in Eastman (1908). More recent finds are summarized in Daeschler and Cressler (2011).

29. Othniel C. Marsh (1896) long ago described what he thought was a Devonian amphibian footprint from Pennsylvania, which he named *Thinopus antiquus*. At first this fossil was accepted as a footprint, but the most recent work on this enigmatic fossil has concluded that it is instead an impression of a group of fish coprolites and not a footprint (Lucas, 2015). Good summaries of the presently accepted European Devonian footprint localities can be found in Lucas (2015, 2019).

30. A rather detailed account of this late Paleozoic glacial age can be found in Eyles and Young (1994). A good summary of this entire interval of ice age conditions can be found at: https://en.wikipedia.org/wiki/Late_Paleozoic_icehouse.

Chapter 8

31. These recent discoveries of amphibians in the Pennsylvania region have been documented by Lucas (2019).

32. One of these earliest reptile trackways was described as *Dromopus adun-cus* by Branson (1910). The other (*Pseudobradypus* isp.) was described from Pennsylvania (Lucas, 2019).

33. The mined Mississippian coal bed lies in the Price Formation. The most recent study of it is by Hower et al., (2008).

Chapter 9

34. Most of the coals that formed in the Pennsylvanian Period occur in the United States and in Western Europe, especially Britain. The greatest concentration and earliest major exploitation of coals in America were in the state of Pennsylvania, from which the American coal-age name "Pennsylvanian" is derived. Pennsylvanian coals are generally found in a series of successive thick beds, whose locations were determined by a series of glacial cycles that occurred during the Pennsylvanian due to large glacial events that were occurring in the continent of Gondwana, which was located in the southern polar region at that time, and Siberia, which like today was located near the northern polar region (see Figure 25).

35. *Arthropleura* is the largest known of these giant Pennsylvanian milli-pedes. It grew to 2.5 meters (8 feet) in length. Tracks probably made by this animal are known from Nova Scotia and are on exhibit at the Natural History Museum in Halifax. An interesting and informative article on these animals can be found at: https://en.wikipedia.org/wiki/Arthropleura.

36. The largest of these was *Meganeura*, but there were other genera alive then that were nearly as large. Although the high oxygen levels at the time they existed have been most commonly suggested as the reason for their large size, it is also quite possible that in the Pennsylvanian there were simply not any vertebrates that were yet adapted to catch and eat these very large flying insects. For information see: https://en.wikipedia.org/wiki/Meganeura.

37. The story of the cyclic pattern of coal deposition and its likely glacial age correlation has been well summarized in the paper by Greb et al. (2008).

Chapter 10

38. A good summary of the factors that are thought to have caused the two great Permian extinction events can be found at: https://en.wikipedia.org/wiki/Permian%E2%80%93Triassic_extinction_event.

39. A more detailed history of this sequence of events can be found in Lucas (2019).

40. A more detailed history of this event can be found in Huang et al. (2016).

41. A more detailed history of this event can be found in Saunders and Reichow (2009) and in Reichow et al. (2009).

42. This occurrence of reworked Permian pollen in the Triassic strata of the Taylorsville Basin was reported by Robbins and Weems (1988). In the Late Triassic, there apparently was a source of Middle or Late Permian pollen that was close to Ashland, Virginia. This pollen was washed out of local rocks and redeposited in the Late Triassic strata of the Taylorsville Basin. It remains unknown whether part of this Permian deposit still remains buried beneath the Late Triassic beds within the Taylorsville Basin or whether this deposit has been totally eroded away.

43. Today, only two species of coelacanth fishes are known to still survive in the deep waters of the Indian Ocean. *Latimeria chalumnae* is found in the Indian Ocean near Africa, while *Latimeria menadoensis* occurs in Indonesia. For more information, see: https://en.wikipedia.org/wiki/Latimeria.

44. The age of the uppermost beds still present in Appalachia has been debated for some time. The most recent and definitive work clearly shows that the uppermost beds preserved in the Appalachian region are Permian and not Pennsylvanian in age (Lucas, 2013; Schneider et al., 2013).

45. A useful summary of the terrestrial vertebrate horizons exposed in Russia can be found in Tverdokhlebov et al. (2005). South African vertebrate beds from a footprint perspective are discussed in Marchetti et al. (2019).

46. A good summary of the global history of Late Permian vertebrates can be found in Bernardi et al. (2017).

APPENDIX 2
SOURCES FOR FIGURES

1. Map of Virginia's geologic regions generated by author.

2. Adapted from the British Geological Survey Geological Time Chart (2022).

3. Locations of Virginia and continental masses during the Cryogenian Period, about 750 Ma. Data largely derived from Wikipedia entry on Rodinia: https://en.wikipedia.org/wiki/Rodinia. Locality data compiled by author. Base map derived from the Appalachian Mountains Joint Venture (AMJV): https://amjv.org/about/.

4. Location of Cryogenian glacial deposits in Virginia, picture generated by author.

5. Photograph of the glacial beds of the Neoproterozoic Konnarock Formation taken by Callan Bentley (1 December 2010), shown at: https://blogs.agu.org/mountainbeltway/2010/12/01/the-konnarock-formation/.

6. Locality data compiled by author. Base map derived from the Appalachian Mountains Joint Venture (AMJV) at: https://amjv.org/about/.

7. *Pteridinium carolinaensis* specimen from Russia (figure from Wikimedia Commons, picture by Ghedoghedo); *Swartpuntia* reconstruction from Narbonne et al. (2015).

8. Location of Virginia during the Ediacaran Period is derived by author from: https://eukaryote.fandom.com/wiki/Ediacaran_Period.

9. Image of the trilobite *Olenellus thompsoni* is from Fenton and Fenton (1958); image of the brachiopod *Nisusia festinata* is from Walcott (1912).

10. Locality data compiled by author. Base map derived from the Appalachian Mountains Joint Venture (AMJV) at: https://amjv.org/about/.

11. Drawing of *Metaspriggina walcotti* by Marianne Collins (2013).

12. Adapted from an illustration by the University of New England (Australia) shown in an online article: https://phys.org/news/2019-08-micro-fossils-extreme-global-environmental.html.

13. Late Cambrian global restoration adapted from: https://www.britannica.com/science/Cambrian-Period

14. Restoration of the oldest well known Ordovician true fossil fish from North America, *Astraspis desiderata*, created by Nabu Tamura (2014).

15. Picture of *Pterygotus* pursuing *Birkenia* fish taken from Wikimedia Commons file: Pterygotus abelov.jpg, which is an image created by ABelov (2014); *Pterygotus* image contains one body segment too many.

16. Picture adapted from "Atlas of Silurian and Middle-Late Ordovician Paleogeographic Maps (Mollweide Projection)," volume 5, map 74, by Christopher Robert Scotese in The Early Paleozoic PALEOMAP Project, Evanston, Ill.

17. Locality data compiled by author. Base map derived from the Appalachian Mountains Joint Venture (AMJV) at: https://amjv.org/about/.

18. Restoration of *Tremataspis* is from Weems and Grimsley (2022); *Poraspis* restoration is from the Wikimedia Commons file: Poraspis.jpg, based on a picture by Lycaon.cl (2009); *Thelodus* restoration is from "Prehistoric Sharks": https://www.sharkattacks.com/prehistoric.htm.

19. Locality data compiled by author. Base map derived from the Appalachian Mountains Joint Venture (AMJV) at: https://amjv.org/about/.

20. *Coccosteus* is from Extinct Animals Wiki: https://prehistopedia.fandom.com/wiki/Coccosteus; *Machaeracanthus* is adapted from Agencia Sinc (2013); *Ctenacanthus* is adapted from Devonian Times: devoniantimes.org/who/pages/ctenacanthus.html; *Limnomis* is adapted from Devonian Times: devoniantimes.org/who/pages/Limnomis.html; *Osteolepis* is adapted from a Wikimedia Commons file created by Edwin Ray Lankester (1908): https://en.m.wikipedia.org/wiki/File:Osteolepis_macro-lepidotus_Guide_to_the_Gallery_of_Fishes.jpg.

21. Reconstructions of *Bothriolepis virginiensis* are from Weems et al. (1981).

22. Image of a Middle Devonian forest landscape created by Zdenek Burian (1963).

23. Restoration of *Hynerpeton bassetti* adapted from "Invasion of the land by tetrapods" by Dmitry Bogdanov (2007); restoration of *Pederpes* created by Dmitry Bogdanov (dmitrchel@mail.ru) (2007).

24. Locality data compiled by author. Base map derived from the Appalachian Mountains Joint Venture (AMJV) at: https://amjv.org/about/.

25. Image adapted from Charles P. T. O'Dale website: https://craterexplorer.ca/mississippian/.

26. *Gyracanthus* image is derived from the Devonian Times: devoniantimes.org/who/pages/gyracanthus.html; *Stethacanthus* image is adapted from Wikimedia Commons file: https://commons.wikimedia.org/wiki/File:StethacanthusesDB_2.jpg by Dmitry Bogdanov (2013); *Tanypterichthys* is modified from Weems and Windolph (1986); *Bluefieldius* is modified from Mickle (2018); *Hadronector* is modified from an online image published by: http://www.fossilmall.com/EDCOPE_Enterprises/fish/fishfossil15/coelacanth.jpg.

27. *Greererpeton* is adapted from a Wikimedia Commons file created by Nobu Tamura (2007): https://commons.wikimedia.org/wiki/File:Greererpeton_BW.jpg; *Proterogyrinus* is adapted from a Wikimedia Commons file created by Dmitry Bogdanov (2007): https://commons.wikimedia.org/wiki/File:Proterogyrinus_DB.jpg.

28. Image of likely *Palaeosauropus* trackmaker is from Lucas (2019); picture was drawn by Matt Celeskey. Footprint images are drawn from Figure 20 in Lucas (2019), except for image of *Pseudobradypus* which is from Mossman and Grantham (1999).

29. Locality data compiled by author. Base map derived from the Appalachian Mountains Joint Venture (AMJV) at: https://amjv.org/about/.

30. *Bandringa* is from an illustration created by Donald Baird (1978); reconstruction of *Janassa* is by Ghedo (2021), done for the Museum für Naturkunde und Vorgeschichte in Dessau; *Orthacanthus* is by Meghunter99, which is posted in "Orthacanthus, Xenacanthiformes, Shark evolution": https://fossil.fandom.com/wiki/Orthacanthus?file=Orthacanthus_1.jpg; *Haplolepis* is from the Fossil Forum: https://www.thefossilforum.com/index.php?/topic/48871-pretty-little-pennsylvanian-coal-fish-found-today/; *Rhizodopsis* is from Wikimedia Commons: https://commons.wikimedia.org/wiki/File:Rhizodopsis.

jpg; *Onychodus* is from Wikimedia Commons: https://commons.wiki-media.org/wiki/File:OnychodusDB15_flipped.jpg, an image by DiB-gd (2015).

31. *Diceratosaurus* adapted from Smokeybjb (2009) at Wikimedia Commons: https://commons.wikimedia.org/wiki/File:Diceratosaurus.jpg; *Diploceraspis* image adapted from Science Planet website: https://science-planet.com/; *Edops* is adapted from Dmitry Bogdanov (2007) at Wikipedia: https://en.wikipedia.org/wiki/Edops; *Fedexia* is adapted from Nobu Tamura (2011) at Wikipedia: https://en.wikipedia.org/wiki/Fedexia; *Diplocaulus* is adapted from the Paleozoic Era Poster Print at Amazon: https://www.amazon.com/Diplocaulus-salamandroi-des-prehistoric-animal-Paleozoic/dp/B07GSR1PPN; *Sauropleura* is adapted from Smokeybjb (2009) at Wikipedia: https://en.wikipedia.org/wiki/Sauropleura.

32. *Hylonomus* is adapted from "Georgia's Pennsylvanian Plant Fossils": https://www.georgiasfossils.com/5a-georgiarsquos-pennsylva-nian-plant-fossils.html; *Archaeothyris* is adapted from Arthur Weasley (2006) at Wikipedia: https://en.wikipedia.org/wiki/Archaeothyris; *Edaphosaurus* is adapted from a painting by Zdeněk Burian (1942).

33. Image of a Pennsylvanian forest adapted from "Plants of the Carbonif-erous Age" by Bibliographisches Institut (ca. 1890) as shown in Wiki-media Commons: https://commons.wikimedia.org/wiki/File:Meyers_b15_s0272b.jpg; image of *Meganeurites* (a close relative of *Meganeura*) is adapted from an image by Handlirsch (1919) as shown in Wikipe-dia: https://en.wikipedia.org/wiki/Meganeura; image of *Arthropleura* is adapted from an image by Neil Davies published in SciTechDaily: https://scitechdaily.com/giant-millipedes-as-big-as-cars-once-roamed-northern-england-complete-fluke-of-a-discovery/.

34. Pennsylvanian footprint images are derived from Lucas (2019).

35. Locality data compiled by author. Base map derived from the Appala-chian Mountains Joint Venture (AMJV) at: https://amjv.org/about/.

36. Image adapted from "Astrobiology at NASA: Recovering from the end-Permian Mass Extinction": https://astrobiology.nasa.gov/news/recovering-from-the-end-permian-mass-extinction/.

37. *Acanthodes* image by P. Janvier (2001), illustrated in Wikipedia: https://

en.wikipedia.org/wiki/Acanthodii; *Amblypterus* image created by H. Alleyne Nicholson (1876), illustrated in Wikipedia: https://en.wikipedia.org/wiki/Amblypterus; *Ectosteorhachis* image by Machiel Mulder (2009), illustrated in Wikimedia Commons: https://commons.wikimedia.org/wiki/File:Ectosteorhachis.jpg; *Xenacanthus* image by Nobu Tamura (2016), illustrated in Wikipedia: https://en.wikipedia.org/wiki/Xenacanthus.

38. *Diploceraspis* image adapted from Science Planet website: https://scienceplanet.com/; *Megamolgophis* image is adapted from Wikiwand: https://www.wikiwand.com/en/Megamolgophis; *Edops* image is adapted from Dmitry Bogdanov (2007) at Wikipedia: https://en.wikipedia.org/wiki/Edops; *Eryops* image is adapted from Dmitry Bogdanov (2007) at Wikipedia: https://en.wikipedia.org/wiki/Eryops.

39. *Ophiacodon* image is adapted from Nobu Tamura (2007) at Wikimedia Commons: https://commons.wikimedia.org/wiki/File:Ophiacodon_BW.jpg; *Ctenospondylus* image is adapted from Nobu Tamura (2007) at Wikipedia: https://en.wikipedia.org/wiki/Ctenospondylus; *Dimetrodon* image is adapted from Charles R. Knight (1897) at Wikipedia: https://en.wikipedia.org/wiki/File:DimetrodonKnight.jpg.

40. Permian footprint images are adapted from illustrations in Lucas (2019).

41. *Moschorhinus* image is adapted from a drawing by Dmitry Bogdanov (2006) at Wikimedia Commons: https://commons.wikimedia.org/wiki/File:Moschorhinus_DB.jpg; *Vivaxosaurus* is adapted from Dmitry Bogdanov (2006) at Wikipedia: https://en.wikipedia.org/wiki/Vivaxosaurus#/media/File:Dicynodon_trautsholdi_DB.jpg; *Sauractonus* is adapted from Wikiwand: https://www.wikiwand.com/en/Sauroctonus; *Ulemosaurus* is adapted from Dmitry Bogdanov (2021) at Wikimedia Commons: https://commons.wikimedia.org/wiki/File:Ulemosaurus_svijagensis.jpg.

42. *Dvinia* image is adapted from a drawing by Nobu Tamura (2017) at Wikipedia: https://en.wikipedia.org/wiki/Dvinia#/media/File:Dvinia_prima_life_restoration.jpg; *Moradisaurus* image is adapted from "Captorhinidae by Karkemish00" on Deviant Art: https://www.deviantart.com/karkemish00/art/Captorhinidae-309263970; *Claudiosaurus* is adapted from "*Claudiosaurus germaini*, an aquatic reptile from

the Upper Permian of Madagascar" by Nobu Tamura as illustrated at Wikimedia Commons: https://commons.wikimedia.org/wiki/File:-Claudiosaurus_white_background.jpg; *Hovasaurus* image adapted from https://yourblog.in.ua/hovasaurus.html; *Arganaceras vacanti* image adapted from Nobu Tamura at Wikipedia: https://en.wikipedia.org/wiki/Arganaceras#/media/File:Arganaceras_BW.jpg; *Pareiasaurus serridens* image adapted from N. Tuchatymuh (Xiphactinus) at Deviant Art: https://www.deviantart.com/xiphactinus/art/Pareiasaurus-serridens-618267724; *Scutosaurus karpinskii* image adapted from Romano, Manucci, Rubridge, and Van den Brandt at Wikimedia Commons: https://commons.wikimedia.org/wiki/File:Scutosaurus_restoration.jpg; *Elginia mirabilis* image adapted from Nobu Tamura at Wikimedia Commons: https://commons.wikimedia.org/wiki/File:Elginia_BW.jpg; *Protorosaurus* image adapted from Emily Stepp (2020) at Deviant Art: https://www.deviantart.com/emilystepp/art/Archosaur-Art-April-2020-Day-15-Protorosaurus-838054229; *Melosaurus platyrhinos* image adapted from Dmitry Bogdanov (2007) at Wikipedia Commons: https://commons.wikimedia.org/wiki/File:Melosaurus_platyrh12DB.jpg; *Rhinesuchus* image adapted from Dmitry Bogdanov (2007) at Wikimedia Commons: https://commons.wikimedia.org/wiki/File:Rhinesuchus1DB.jpg; *Dicynodontoides* (= *Kingoria*) image adapted from Dmitry Bogdanov (2007) at Wikimedia Commons: https://commons.wikimedia.org/wiki/File:KingoriaDB.jpg.

TABLE 1

VERTEBRATE FOSSILS DESCRIBED FROM THE PALEOZOIC STRATA OF VIRGINIA AND NEARBY STATES

CAMBRIAN (at least 2 biological species of conodonts; 1 vertebrate species)

CONODONTA

Cambrooistodus minutus (Va.)

Clavohamulus elongatus? (Va.)

Cordylodus proavus (Va.)

Eoconodontus notchpeakensis (Va.)

Hirsutodontus hirsutus (Va.)

Phakelodus tenuis (Va.)

Proconodontus muelleri (Va.)

Proconodontus serratus (Va.)

Prooneotodus rotundatus (Va.)

Teridontus nakamuri (Va.)

Westergaardodina sp. (Va.)

CHORDATA

Metaspriggina sp. (Pa.)

ORDOVICIAN (at least 10 biological species of conodonts; 0 vertebrate species)

CONODONTA

"Acanthodus" lineatus (Va.)

"Acanthodus" uncinatus (Va.)

cf. *Acodus oneotensis* (Va.)

"Acontiodus" iowensis (Va.)

"Acontiodus" staufferi (Va.)

Acontiodus propinquus (Va.)

Chosonodina herfurthi (Va.)

Clavohamulus densus (Va.)

Clavohamulus elongatus? (Va.)

Colaptoconus quadraplicatus (Va.)

Cordylodus angulatus (Va.)

Cordylodus intermedius (Va.)

Cordylodus lindstromi (Va.)

Cordylodus proavus (Va.)

Cordylodus rotundatus (Va.)

Curtognathus robustus (Va.)

Drepanodus arcuatus (Va.)

Drepanodus concavus (Va.)

Drepanodus pseudoconcavus (Va.)

Drepanoistodus basiovolis (Va.)

Drepanoistodus pervetus (Va.)

Drepanoistodus suberectus (Va.)

Eoconodontus notchpeakensis (Va.)

Eucharodus parallelus (Va.)

Eucharodus toomeyi (Va.)

Fryxellodontus inornatus (Va.)

Fryxellodontus lineatus (Va.)

Glyptoconus quadraplicatus (Va.)

Hirsutodontus hirsutus (Va.)

Iapetognathus? sp. (Va.)

Juanognathus? n. sp. (Va.)

Loxodus bransoni (Va.)

Monocostodus sp. (Va.)

Oepikodus communis (Va.)

"Oistodus" triangularis (Va.)

"Oistodus" venustus (Va.)

Oneotodus costatus (Va.)

Paltodus bassleri (Va.)

Panderodus gracilis (Va.)

Paroistodus? sp. (Va.)

Plectodina joachimensis (Va.)

Rossodus manitouensis (Va.)

Scandopus? sp. (Va.)

"Scolopodus" filosus (Va.)

"Scolopodus" gracilis (Va.)

"Scolopodus" sulcatus (Va.)

Semiacontiodus nogamii (Va.)

Teridontus nakamurai (Va.)

Tropodus comptus (Va.)

Ulrichodina abnormalis (Va.)

Ulrichodina deflexa (Va.)

Ulrichodina wisconsinensis (Va.)

Utahconus utahensis (Va.)

Variabiloconus bassieri (Va.)

SILURIAN (at least 12 conodont species; 5 vertebrate species)

CONODONTA

Apsidognathus tuberculatus (W. Va.)

Bellodella sp. (Va., W. Va)

Carniodus carnulus (Md., Va., W. Va.)

Delotaxis elegans (Va., W. Va.)

"Distomodus" dubius (W. Va)

Distomodus nodusus (Md., Va., W. Va.) (This is an element set

name including cf. *Trichonodella? expansa,* cf. *Distomodus
kentuckyensis,* cf. *Roundya truncialata,* and cf. *Ligonodina
egregia*)

Hindeodella confluens (Md., Va., W. Va.)

Hindeodella priscilla (part of element set XI) (Md., Va., W. Va.)

Kockelella walliseri (W. Va.)

Ligonodina n. sp. (part of element set III) (Va., W. Va.)

Ligonodina brevis (part of element sets II, VIII, X) (Md., Va., W. Va.)

Ligonodina? silurica (part of element sets IV, VII, IX) (Md., Va., W. Va.)

Lonchodina detorta (Md., Va., W. Va.)

Lonchodina? greilingi (part of element sets II, VIII, X) (Md., Va., W. Va.)

Lonchodina walliseri (part of element set V) (Md., Va., W. Va.)

Neoprioniodus excavatus (part of element sets IV, VII, IX) (Md., Va., W. Va.)

Neoprioniodus multiformis (Va.)

Oulodus confluens (W. Va)

Oulodus aff. *O. cristagalli* (W. Va)

Oulodus elegans (Va., W. Va)

Ozarkodina aequalis (part of element set V) (Va., W. Va.)

Ozarkodina confluens (part of element set VI) (W. Va.)

Ozarkodina edithae (part of element set I) (Va., W. Va.)

Ozarkodina excavata (Md., Va., W. Va)

Ozarkodina gulletensis (Md., Va., W. Va.) (This is an element set
name including *Spathognathodus gulletensis, Ozarkodina al-
lisonae, Prioniodus bicurvatus, Trichognathus symmetrica, Synprioniodina silurica,* and *Hindeodella equidentata*)

Ozarkodina ortuformis (part of element set X) (Md., Va., W. Va.)

Ozarkodina remscheidensis (Va., W. Va.)

Ozarkodina serrata (part of element set III) (Va., W. Va.)

Ozarkodina sinuosa (part of element set V) (Md., Va., W. Va.)

Ozarkodina snajdri (part of element set III) (Va., W. Va)

Ozarkodina steinhornensis (Va., W. Va.)

Ozarkodina tenuiramea (part of element set V) (Md., Va., W. Va.)

Ozarkodina typica (part of element sets IV, VI, VII, IX, XI) (Md., Va., W. Va.)

Ozarkodina cf. *O. wimani* (W. Va)

Ozarkodina ziegleri (part of element sets II, VIII) (Md., Va., W. Va.)

Panderodus serratus (Md., Va., W. Va.)

Panderodus simplex (Md., Va., W. Va.)

Panderosus unicostatus (Va., W. Va.)

Plectospathodus? n. sp. (part of element set III) (Va., W. Va.)

Plectospathodus alternatus (part of element set XI) (Md., Va., W. Va.)

Plectospathodus extensus (part of element sets IV, VII, IX) (Md., Va., W. Va.)

Plectospathodus flexuosus (part of element set VI) (Md., Va., W. Va.)

Pseudooneotodus beckmanni (W. Va)

Pseudooneotodus bicornis (W. Va.)

Pterospathodus amorphognathoides (Md., W. Va.)

Pterospathodus celloni (Md., W. Va.)

Spathognathodus bicornutus (part of element set II) (Va., W. Va.)

Spathognathodus crispus (part of element set X) (Md., Va., W. Va.)

Spathognathodus primus (part of element sets IV, VI, VII, IX) (Md., Va., W. Va.)

Spathognathodus sagitta (part of element set I) (Va., W. Va.)

Spathognathodus snajdri (Va., W. Va.)

Spathognathodus steinhornensis (element set XI) (Md., Va., W. Va.)

Spathognathodus tillmani (part of element set VIII) (Md., Va., W. Va.)

Spathognathodus walliseri (Va., W. Va.)

Synprioniodina n. sp. (part of element set III) (Va., W. Va.)

Synprioniodina bicurvata (part of element sets VI, XI) (Md., Va., W. Va.)

Synprioniodina lowryi (part of element sets II, VIII, X) (Md., Va., W. Va.)

Trichonodella n. sp. (part of element set III) (Va., W. Va.)

Trichonodella excavata (part of element sets IV, VII, IX) (Md., Va., W. Va.)

Trichonodella inconstans (part of element sets II, VIII, X) (Md., Va., W. Va.)

Trichonodella symmetrica (part of element sets VI, XI) (Md., Va., W. Va.)

Walliserodus curvatus (Md., W. Va.)

OSTEOSTRACI

aff. *Tremataspis* sp. (Va.)

PTERASPIDOMORPHI

Americaspis americana (Pa.)

Vernonaspis allenae (Md., Pa. W.Va.)

THELODONTOMORPHI

Logania sp. (Pa.)

Thelodus sp. (Pa.)

DEVONIAN (at least 6 conodont species; 34 vertebrate species)

CONODONTA

Amorphognathus tvaerensis (Va.)

Ancyrodella aff. *A. alata* (Va.)

Ansella sp. (Va.)

Appalachignathus delicatulus (Va.)

Baltoniodus gerdae (Va.)

Bellodella sp. (Va.)

Chosonodina herfurthi (Va.)

"Dapsilodus" simularis (Va.)

Decoriconus sp. (Va.)

Diaphorodus delicatus (Va.)

Dvorakia sp. (Va.)

Erismodus asymmetrica (Va.)

Erismodus radicans (Va.)

Eucharodus sp. (Va.)

Icriodus claudiae (Va.)

Icriodus corniger (Va.)

Icriodus symmetricus (Va.)

Icriodus woschmidti (W. Va.)

Juanognathus? sp. (Va.)

Leptochirognathus primus (Va.)

Leptochirognathus quadratus (Va.)

Loxodus bransoni (Va.)

Macerodus dianae (Va.)

Mehlina gradata (Va.)

Mesotaxis asymmetricus (Va.)

Milaculum sp. (Va.)

Neomultioistodus clypeus (Va.)

Oulodus cristagalli (Va.)

Pandorinellina insita (Va.)

Paraprionodus costatus (Va.)

Pedovis sp. (Va.)

Periodon aculeatus (Va.)

Periodon grandis (Va.)

Phragmodus flexuosus (Va.)

Phragmodus undatus (Va.)

Polygnathus cooperi (Va.)

Polygnathus dubius (Va.)

Polygnathus linguiformis (Va.)

Protopanderodus liripipus (Va.)

Pseudooneotodus beckmanni (Va.)

Pseudooneotodus migratus (Va.)

Pteracontiodus aff. *P. cryptodens* (Va.)

Scolopodus floweri (Va.)

Ulrichodina abnormalis (Va.)

Utahconus? sp. (Va.)

PLACODERMI

Bothriolepis nitida (Pa., W. Va.?)

Bothriolepis virginiensis (Va.)

Dinichthys curtus (Pa.)

Dinichthys tuberculatus (Pa.)

Eczematolepis sp. (= *Acantholepis* sp.) (Va.)

Glyptaspis eastmani (Md.)

Groenlandaspis pennsylvanica (Pa.)

Holonema horridum (Pa.)

Holonema rugosum (Md., Pa.)

Phyllolepis delicatula (Pa.)

Phyllolepis rossimontina (Pa.)

Spenophorus lilleyi (Pa.)

Turrisaspis elektor (Pa.)

ACANTHODII

Gyracanthus sherwoodi (Pa.)

Machaeracanthus peracutus (Va.)

CHONDRICHTHYES

Ageleodus pectinatus (Pa.)

Cladodus coniger (Pa.)

Ctenacanthus sp. (Pa.)

Helodus gibberulus (Pa.)

Homacanthus acinaciformis (Pa.)

ACTINOPTERYGII

Limnomis delaneyi (Pa.)

SARCOPTERYGII

Apedodus priscus (Pa.)

Dipterus angustus (Pa.) (*D. contraversus, D. fleischeri, D. minutus, D. nelsoni,* and *D. sherwoodi* may be synonyms of *D. angustus*)

Ganorhynchus beecheri (Pa.)

Ganorhynchus oblongus (Pa.)

Glypotopomus sayrei (Pa.)

Heliodus lesleyi (Pa.)

Holoptychius americanus sp. (Pa., W. Va.?) (*H. filosus, H. giganteus, H. flabellatus, H. latus, H. radiatus,* and *H. serrulatus* may be synonyms with *H. americanus*)

Hyneria lindae (Pa.)

Osteolepis sp. (Pa.)

Sauripterus taylori (Pa.)

AMPHIBIA

Hynerpeton bassetti (Pa.)

Densignathus rowei (Pa.)

Whatcheeridae gen. et sp. indet. (Pa.)

CARBONIFEROUS (MISSISSIPPIAN) (at least 2 conodont species, 23 vertebrate species)

CONODONTA

Adetognathus unicornis (W. Va.)

Cavusgnathus convexus (W. Va.)

Cavusgnathus unicornis (Md., Pa.)

Gnathodus bilineatus (W. Va.)

Gnathodus texanus? (Pa.)

Hindeodus cristulus (Md.)

Hindeodus spiculus (Md.)

Kladognathus sp. (Md., Pa.)

Lochriea commutata (Md.)

Polygnathus sp. (Pa.)

Vogelgnathus campbelli? (Md.)

ACANTHODII

Gyracanthus sp. (W. Va.)

CHONDRICHTHYES

Cladodus sp. (Pa.)

Desmiodus tumidus (Pa.)

Falcatus falcatus (W. Va.)

aff. *Helodus* sp. (W. Va.)

Hybocladodus sp. (Pa.)

Poecilodus st. ludovicii (W. Va.)

Psephodus concolutus (Pa.)

Psephodus crenulatus (Pa.)

Venustodus argutus (Pa.)

Venustodus leidyi (Pa.)

Venustodus variabilis (Pa.)

ACTINOPTERYGII

Bluefieldius mercerensis (W. Va)

Tanypterichthys pridensis (W.Va.)

SARCOPTERYGII

Tranodis castrensis (W. Va.)

AMPHIBIA

Greererpeton burkemorani (W.Va.)

Proterogyrinus scheelei (W.Va.)

Batrachichnus salamandroides (Pa.) (ichnotaxon)

Hylopus hardingi (Pa.) (ichnotaxon)

Matthewichnus isp. (Pa.) (ichnotaxon)

Palaeosauropus primaevus (Pa.) (ichnotaxon)

REPTILIA

Dromopus aduncus (Va.) (ichnotaxon)

Pseudobradypus isp. (Pa.) (ichnotaxon)

CARBONIFEROUS (PENNSYLVANIAN) (at least 1 conodont species; 69 vertebrate species)

CONODONTA

Idiognathodus cancellosus (Pa.)

Idiognathodus cherryvalensis? (Pa.)

Idiognathodus confragus (Pa. W. Va.)

Idiognathodus swadei (W. Va.)

Streptognathodus corrugatus (Pa.)

Streptognathodus excelsus (Pa.)

CHONDRICHTHYES

Agassizodus variabilis (Pa.)

Ageleodus sp. (W. Va.)

Bandringa rayi (Pa.)

Chomatodus sp. (W. Va.)

Cladodus occidentalis (Pa., W. Va.)

Deltodus angularis (Pa., W. Va.)

Dittodus sp. (W. Pa.)

Fissodus inaequalis (Pa.)

Helodus simplex (W. Va.)

Janassa striglina (W. Va.)

Orthacanthus arcuatus (Oh., W. Va.?)

"Peltodus" transversus (W. Va.)

Peripristis semicircularis (W. Va.)

Petalodus ohioensis (Pa., W. Va.)

Physonemus cf. *P. acinaciformis* (W. Va.)

Xenacanthus compressus (Oh., Pa.)

Xenacanthus gracilis (Oh.)

Xenacanthus cf. *X. triodus* (W. Va.)

ACTINOPTERYGII

aff. *Commentrya* sp. (Pa.)

Haplolepis corrugata (Oh.)

Haplolepis aff. *H. ovoidea* (Pa.)

Elonichthys peltigerus (Oh., Pa.)

Microhaplolepis ovoidea (Oh.)

Microhaplolepis serrata (Oh.)

Parahaplolepis tuberculata (Oh.)

Pyritocephalus lineatus (Oh.)

aff. *Sphaerolepis* sp. (W. Va.)

SARCOPTERYGII

Conchopoma exanthematicum (Oh.)

cf. *Ectosteorhachis nitidus* (W. Va.)

Megalichthys sp. (W. Va.)

Onychodus sp. (W. Va.)

Rhabdoderma elegans (Oh., Pa.)

Rhizodopsis cf. *R. robustus* (Pa.)

Sagenodus cf. *periprion* (W. Va.)

Sagenodus serratus (Oh., W. Va.)

AMPHIBIA

Amphibamus lyelli (Oh.)

Brachydectes newberryi (Oh., W. Va.)

Colosteus scutellatus (Oh.)

Ctenerpeton remex (Oh.)

Desmatodon hollandi (Pa.)

Diceratosaurus brevirostris (Oh.)

Diplocaulus salamandroides (W. Va.)

Diploceraspis burkei (= *D. conemaughensis*) (W. Va.)

Edops sp. (W. Va.)

Erpetosaurus radiatus (Oh.)

Eusauropleura digitata (Oh.)

Fedexia striegeli (Pa.)

Gandrya cf. *G. latistoma* (Oh.)

Glaucerpeton avinoffi (Pa., W.Va.)

Isodectes obtusus (Pa.)

Leptophractus obsoletus (Oh.)

Lysorophus dunkardensis (Pa., W. Va.)

Macrerpeton huxleyi (Oh.)

Megalocephalus enchodus (Oh.)

Molgophis macrurus (Oh.)

Odonterpeton triangulare (Oh.)

Oestocephalus amphiuminum (Oh.)

Phlegethontia linearis (Oh.)

Pleuroptyx clavatus (Oh.)

Ptyonius marshii (Oh.)

Raphetes lineolatus (Oh.)

Saurerpeton obtusum (Oh., W. Va.)

Sauropleura pectinata (Oh.)

Stegops newberryi (Oh.)

Tuditanus punctulatus (Oh.)

Batrachichnus salamandroides (Va.) (ichnotaxon)

Characichnos isp. (Va.) (ichnotaxon)

Limnopus heterodactylus (Va.) (ichnotaxon)

Limnopus isp. (Va.) (ichnotaxon)

REPTILIA
Anthracodromeus longipes (Oh.)
Cephalerpeton aff. *C. ventriarmatum* (Oh.)

SYNAPSIDA
Archaeothyris sp. (Oh.)
Edaphosaurus colohistion (W. Va.)

PERMIAN (at least 1 conodont species; 27 vertebrate species)

CONODONTA
Hindeodus calcarus (Oh.)
Idiognathodus podolskensis (Oh.)
Idioprioniodus conjunctus (Oh.)

ACANTHODII
Acanthodes marshi (W. Va.)

CHONDRICHTHYES
Dittodus sp. (Pa.)
Hybodus allegheniensis (Pa.)
Orthacanthus sp. (Pa.)
Xenacanthus sp. (Pa.)

ACTINOPTERYGII
Amblypterus stewarti (W. Va.)
Palaeoniscoidea indet. (W. Va., Pa.)

CROSSOPTERYGII
cf. *Ectosteorhachis nitidus* (Pa., W. Va.)
Monongahela sp. (Pa.)
Sagenodus cf. *periprion* (Pa., W. Va.)
Sagenodus serratus (Oh., W. Va.)

AMPHIBIA
Ambedus pusillus (Oh.)

Brachydectes sp. (= holotype of *Lysorophus minutus*, nomen dubium) (W. Va.)

Diadectes sp. (W. Va.)

Diploceraspis burkei (Oh., Pa., W. Va.)

aff. *Edops* sp. (W. Va.)

Eryops cf. *E. megacephalus* (Pa., W. Va.)

Lysorophus dunkardensis (Pa., W. Va.)

Megamolgophis agostinii (Pa., W. Va.)

Phlegethontia sp. (Oh.)

Baropus waynesburgensis (ichnotaxon of a rhachitomous amphibian) (W. Va.)

Batrachichnus isp. (ichnotaxon of a temnospondyl or microsaur amphibian) (Md.)

Limnopus isp. (ichnotaxon likely made by *Diadectes*) (W.Va.)

REPTILIA

Protorothyris morani (W. Va.)

SYNAPSIDA

Archaeothyris sp. (Oh.)

Ctenospondylus sp. (Oh.)

Dimetrodon sp. (Oh., Pa.)

Edaphosaurus cf. *E. boanerges* (Pa., W. Va.)

Edaphosaurus cf. *E. cruciger* (W. Va.)

Ophiacodon sp. (Oh.)

Dimetropus berea (ichnotaxon, probably of *Edaphosaurus*) (Oh., Pa., W. Va.)

Dromopus sp. (ichnotaxon, likely of an araeoscelid or a bolosaurid) (Md., Oh., Pa., W. Va.)

ABOUT THE AUTHORS

Robert E. Weems grew up in Ashland, Virginia. In the second grade, he read a book called All About Dinosaurs by Roy Chapman Andrews and from then on became determined to learn all he could about the fossil animals and plants in Virginia and nearby states. He earned a bachelor's degree in biology from Randolph-Macon College in 1968, a master's degree in geology from Virginia Tech in 1972 and a doctoral degree from the George Washington University in 1978. Dr. Weems worked for the U. S. Geological Survey in Reston, Virginia, from 1978 to 2010 studying and mapping the stratigraphy of piedmont and coastal plain strata in the eastern United States. Since then, he has spent much of his retirement continuing research on the fossil animals and plants in and around the Virginia region.

Gary J. Grimsley grew up in Annandale, Virginia. Like his co-author, he became fascinated with fossils and paleontology at a young age upon reading several books on dinosaurs and other animals from the past, especially Fossils (A Golden Guide) by Paul R. Shaffer, Frank H. T. Rhodes and Herbert S. Zim,

which he read and reread many times. He earned a bachelor's degree in business administration from Virginia Tech in 1977. Since then, he has spent decades collecting and studying fossils in the Mid-Atlantic region and for seven years was a member of the board of directors for the Maryland Geological Society. He is currently collaborating with several professional paleontologists on research projects and publications about the Maryland and Virginia regions.

www.ingramcontent.com/pod-product-compliance
Lightning Source LLC
Chambersburg PA
CBHW071228290326
41931CB00037B/2453